Operations Management - Recent Advances and New Perspectives

Edited by Tamás Bányai

Published in London, United Kingdom

Operations Management - Recent Advances and New Perspectives
http://dx.doi.org/10.5772/intechopen.111296
Edited by Tamás Bányai

Contributors
Ahmed Esmail Mohamed, Andrés Felipe Santos, Driss Essabbar, Marc Zolghadri, Maria Zrikem, Marvel Ogah, Natalia Camacho Franco, Olumide Odeyinka, Olumide Omoegun, Vladimir Križaić

First published in London, United Kingdom, 2024 by IntechOpen
IntechOpen is the global imprint of INTECHOPEN LIMITED, registered in England and Wales, registration number: 11086078, 167-169 Great Portland Street, London, W1W 5PF, United Kingdom

British Library Cataloguing-in-Publication Data
A catalogue record for this book is available from the British Library

Additional hard and PDF copies can be obtained from orders@intechopen.com

Operations Management - Recent Advances and New Perspectives
Edited by Tamás Bányai
p. cm.

This title is part of the Industrial Engineering and Management Book Series, Volume 5
Topic: Production Engineering
Series Editor: Fausto Pedro Garcia Marquez
Topic Editor: Orhan Korhan

Print ISBN 978-0-85014-156-6
Online ISBN 978-0-85014-157-3
eBook (PDF) ISBN 978-0-85014-158-0
ISSN 3029-0511

For EU product safety concerns:
IN TECH d.o.o., Prolaz Marije Krucifikse Kozulić 3, 51000 Rijeka, Croatia,
info@intechopen.com or visit our website at intechopen.com.

IntechOpen Book Series

Industrial Engineering and Management

Volume 5

Aims and Scope of the Series

Industrial Engineering and Management (IEM) is a discipline that focuses on optimizing complex processes and systems within various industries. It involves the integration of engineering, business, economics, mathematics, and behavioral sciences to improve efficiency, productivity, quality, and overall performance in organizations. Key aspects of Industrial Engineering and Management include: Process Optimization; System Analysis and Design; Quality Control and Management; Supply Chain Management; Operations Management; Human Factors and Ergonomics; Project Management; Cost Analysis and Financial Management; Decision Analysis.

Overall, Industrial Engineering and Management aims to optimize resources, improve processes, enhance productivity, and ensure the effective and efficient utilization of all elements involved in the production or delivery of goods and services. It is crucial in today's competitive business environment for organizations to stay efficient and competitive.

Production Engineering and Operational Excellence are fields of study and practices that focus on optimizing and improving the manufacturing and production processes within an organization. It combines principles from engineering, management, and operational strategies to enhance productivity, efficiency, quality, safety, and sustainability in the production of goods and services.

Here are the key components of Production Engineering and Operational Excellence: Process Optimization; Operational Excellence; Manufacturing Systems Design; Quality Management; Supply Chain Optimization; Production Planning and Scheduling; Automation and Technology Integration; Health, Safety, and Environmental Management; Cost Management; Performance Measurement and Key Performance Indicators (KPIs); Continuous Improvement and Innovation.

Production Engineering and Operational Excellence are crucial for organizations aiming to stay competitive in the global market by achieving high levels of efficiency, quality, and customer satisfaction while optimizing resources and minimizing waste. It is a multidisciplinary approach that encompasses engineering principles, management strategies, and the effective use of technology to drive operational success.

Meet the Series Editor

Fausto Pedro Garcia Marquez is a Full Professor at UCLM, Spain, with accreditation since 2013. He also holds the position of Honorary Senior Research Fellow at Birmingham University, UK, and serves as a Lecturer at the Postgraduate European Institute. In addition to these roles, Fausto has experience as a Senior Manager at Accenture from 2013 to 2014. He earned his European Ph.D. with the highest distinction. Throughout his career, Fausto has received numerous awards and honors. These include the Nominate Prize (2022), Gran Maestre (2022), Grand Prize (2021), Runner Prize (2020), and Advancement Prize (2018), as well as Runner (2015), Advancement (2013), and Silver (2012) by the International Society of Management Science and Engineering Management (ICMSEM). He was also the recipient of the First International Business Ideas Competition 2017 Award. Fausto's contributions extend to academic publishing, with over 242 papers to his name. Notably, his work has been recognized in journals like "Applied Energy" (Q1, IF 9.746, Best Paper 2020) and "Renewable Energy" (Q1, IF 8.001, Best Paper 2014). His affiliations include the editorial and authorship roles in more than 50 books, with publications through respected publishers such as Elsevier, Springer, Pearson, Mc-GrawHill, IntechOpen, IGI, Marcombo, and AlfaOmega. He has authored over 100 international chapters and holds 6 patents. Fausto serves as the Editor of 5 International Journals and is a Committee Member for more than 70 International Conferences. His research portfolio encompasses being the Principal Investigator in 4 European Projects, 8 National Projects, and participating in over 150 projects involving universities and companies. His areas of expertise and research interests span Artificial Intelligence, Maintenance, Management, Renewable Energy, Transport, Advanced Analytics, and Data Science. Fausto is a recognized Expert in the European Union in AI4People (EISMD) and ESF. He also serves as the Director of www.ingeniumgroup.eu, holds the status of Senior Member at IEEE since 2021, and has been honored as an Honorary Member of the Research Council of the Indian Institute of Finance since 2021. Fausto is also the Committee Chair of The International Society for Management Science and Engineering Management (ISMSEM) since 2020.

Meet the Volume Editor

Tamás Bányai received a master's degree in 1993 and a Ph.D. in 1999, both from the University of Miskolc, Hungary, where he is currently a full-time professor. He has 30 years of teaching and research experience in the design and control of materials handling systems and supply chain management, with special emphasis on heuristic optimization of large-scale systems. He has published more than 200 research papers, book chapters, and conference proceedings. He has been a member and manager of more than 50 national and international R&D projects. Away from academia, Prof. Bányai's other interests include playing the piano and taking photographs.

Contents

Preface

This book offers a selection of chapters on operation management, promoting new research results in the field. The book covers six topics, determined by the theoretical and practical aspects of operation management.

Chapter 1, "An Analysis Framework of Power Imbalance in Supply Chains", focuses on the analysis of power in networks of supply chain partners. This chapter proposes a novel approach to the power analysis framework by presenting a set of concepts and properties. The proposed framework makes it possible to model and examine various aspects of power and the interactive process in networks of supply chain partners.

Chapter 2, "Driving Operational Resilience: Perspectives from Emerging and Developing Economies", focuses on supply chain resilience and incidents of supply chain disruptions. Global best practices for eliciting operational resilience and key recommendations for key stakeholders are shown. The chapter focuses on strategies to mitigate and build resilience for developing countries, while recommendations for policy-makers and key stakeholders are also proposed.

Chapter 3, "Inventory Management", focuses on inventory management, emphasizing inventory management objectives, techniques, and best practices. The chapter discusses vendor-managed inventory (VMI) and just-in-time (JIT) and highlights the importance of accurate forecasting, which makes it possible to reduce carrying costs, enhance cash flow, and streamline supply chain operations.

Chapter 4, "Emerging Inventory Planning", demonstrates the importance of appropriate inventory management strategies. The chapter focuses on the impact of uncertain production and service environments and shows that supply chain members have chosen to stockpile large quantities of inventory, increasing working capital while trying to reduce return on assets (ROA). The authors show the importance of logistics planning and strategies to support a responsive inventory.

Chapter 5, "MGSC Project Process Management", proposes a novel approach for financial flow forecasting, which is determined by the modification of the Gaussian S-curve (MGSC) through scientific project management. A dynamic simulation validates the proposed methodology.

Chapter 6, "Warehouse Operations: An Examination of Traditional and Automated Approaches in Supply Chain Management", focuses on the different approaches to warehouse operations and discusses the need for efficient warehouse management to enhance customer satisfaction, reduce costs, and improve overall productivity. The importance of an appropriate warehouse management system (WMS) is also shown.

The aim of this book is to help students as well as managers and researchers to understand and appreciate the concept, design, and implementation of operation management. The editors thank the chapter authors for their scientific contributions. The chapters were edited and published following a rigorous selection process. We also wish to thank and acknowledge the many individuals who helped us throughout the editorial process that made this book possible.

Tamás Bányai
Institute of Logistics,
University of Miskolc,
Miskolc, Hungary

Chapter 1

An Analysis Framework of Power Imbalance in Supply Chains

Driss Essabbar, Maria Zrikem and Marc Zolghadri

Abstract

Is a huge multinational company always powerful, and is a small company always weak? What actually makes a company stronger than another one? These are two typical questions students, researchers and practitioners encounter when they study or practice management. Decision-makers in companies have to be able to account for power in their daily tasks because of the potential partners' reactions to wrong decisions. Major papers studying power do agree that power and its direct consequence, i.e., decision imbalance, could have direct impacts on how soundly a supply chain operates. Even though power has been analyzed for a long time in various fields, to the best of our knowledge, there is a lack of scientific literature examining and modeling power imbalance and its consequences in supply chains. An important challenge regarding this issue is analyzing power in networks of partners. This chapter contributes to the development of a power analysis framework by presenting a set of concepts and properties to model and examine various aspects of power and the interactive process that leads to its use or does not. This framework will use data collected from a set of interviews conducted with main decision-makers to analyze a real case of four companies from the high-tech sector. The chapter does not address strategies or techniques to cope with power exertion but helps to elucidate the process of power evolution.

Keywords: power, supply chain, collaboration, influence analysis, partnership

1. Introduction

Modern companies focus on their core businesses to enter new markets or to consolidate their advantages. Therefore, manufacturing products may require partner coordination. There are several strategic suppliers sharing risks and revenues with which the focal company tends to create long-term and stable relationships. This association is known as an enterprise network or supply chain [1–4]. These partnerships correspond to links between independent firms [5] seeking to achieve goals that each one could not attain when acting alone [6]. Collaboration should be based on mutual trust and commitment by sharing a common vision of the supply chain's objectives. A number of studies [2, 7–11] support that developing collaborative relationships has a positive influence on business success. However, this target is quite difficult to achieve, given the opportunistic behavior of certain partners who might intend to take advantage of the established relations. These companies give priority to their own objectives first and could deteriorate the whole supply chain's performance (see for instance Tost [12]).

The development of relationships between companies is significantly affected by a key factor, which is the power imbalance. This imbalance primarily arises when a more powerful company can impose unfavourable terms on its less powerful counterparts. In a broader sense, power can be defined as the degree to which individuals believe they can exert control over the results and experiences of others; see Fast et al. [13]. This concept will be discussed in further detail later in this chapter. Being able to apply pressure upon weaker actors means that the stronger partner has been able to understand its real source of power and was able to use it. However, if a company overestimates or underestimates its own power, its consequent actions could affect its ability to negotiate successful collaboration with the supplier or customer and reduce its agility [14]. The role of power is crucial to its interactions with other relationship structures; power can seriously affect the level of confidence and commitment between parties [15].

Several authors [16, 17] have discussed that the exertion of power, in the case of buyers and suppliers for instance, is likely to result in unproductive partnerships because, under such circumstances, exploitation, rather than cooperation, might result [15]. Zolkiewski [18] emphasized that power plays a central role in networks, influencing the contest for resource control, shaping the networks' overall dynamics and potentially giving rise to competitive dynamics. To reinforce this idea, Cendon and Jarvenpaa [19] have identified the direct impact of a power imbalance on strategic choices, governance structures, relative dependence, resources and activities performed by a supply chain, while [15] suggest that the power of network actors is likely to influence the distribution of responsibilities and the flow of benefits between them significantly according to their dependencies.

According to Porter [20], the market position of a firm is the main focus of strategy building. In this respect, his five forces model also involves a struggle for control over buyers and suppliers (parties perceived as representing threats to the success of any focal firm). The factors at play here encompass the bargaining strength of suppliers, the potential for substitute products, the threat of new entrants, rivalry among competitors and the bargaining power held by customers. According to Trienekens and Beulens [21], companies should base their strategy on the status and development of these forces.

One of the key tangible reasons for conducting research on power is to proactively mitigate potential conflicts and operational disruptions that might arise in the product design phase or due to implicit imbalances in decision-making within a supply chain. Analyzing power dynamics offers the benefit of contextualizing a company within a web of interdependencies. It prompts decision-makers to view their company as a node within a broader network upon which they rely. **Table 1** summarizes some relevant papers from the scientific literature on power where authors show how these dependencies influence a number of activities in the supply chain, such as the choice of strategies (partner choice, market development), coordination planning, risk management [29] or global performance. Hereafter, we focus on two typical cases among the industrial-based cases found within the scientific literature.

The development of Audi's and Suzuki's supply chains in Hungary was studied by Demeter et al. [30] through interviews with supply chain managers of 13 different companies including two assemblers (Audi and Suzuki), two system integrators, and nine direct and indirect suppliers. In the case of Audi, the purchased volume of each part is divided between two suppliers, and they cannot obtain more than 30–50% of

Authors	Use case	Conclusion
Ming et al. [22]	Aeronautical supply chains	Trust, power, commitment and dependence are some of the non-technical concepts that may help to develop an understanding of a situation of cooperation and its underlying problems.
Shou et al.	Industrial sectors in the Yangtze River Delta	Innovative, financial, human and organizational resources are significantly important for the companies to take control over others. Organizational resources help the buyer to create its power over suppliers.
Belaya and Hanf [23]	The relationships between suppliers and buyers in Russian agri-food supply chains	Power can be viewed as one of the most accessible and potent instruments in vertical relationships throughout the entire food supply chain. Every participant in the supply chain endeavours to establish a dominant position, enabling them to exert control over other members and maximize their earnings through the exercise of their power.
Munksgaard et al. [24]	A multiple case study of companies in the food industry, the construction industry and the energy sector	The research affirms that leading companies, whether categorized as hubs, brokers or facilitators, can either enhance or hinder the collaborative development initiatives within the network.
Wei [25]	Case studies in two industries (automotive industry and laptop computer industry)	Power seems to have an impact on schedule flexibility, either through direct or indirect means. The supplier's ability to adapt their processes can diminish the effect of customer power, whereas the customer's willingness to share information and their commitment to purchase quantity may not necessarily be contingent on their level of power.
Su et al. [4]	The Japanese sugar dispute	An actor's positional power is also normative in that actors can also exert influence over others through shared values and expectations. Relationship atmosphere and indirect power were perceived as crucial to understanding sources of conflict and conflict management, respectively.
Zhuang et al. [26]	The research was conducted in the People's Republic of China. The dyadic relationship examined is between department stores and their suppliers	The study found that channel members' power (supply chain members) has a positive impact on their uses of non-coercive power regardless of their relative position in channel dyads; their uses of coercive power are negatively sensitive to their own power if they are at a power advantage (stronger), and positively sensitive to the target members' power if they are at a power disadvantage (weaker).
Benton and Maloni [16]	Daimler-Chrysler (with 178 suppliers) and Honda (with 392 suppliers)	These findings underscore the significance of skilful power management as a strategy for improving supply chain integration. This, in turn, fosters a heightened awareness of the influential role of power in the supply chain,

Authors	Use case	Conclusion
		which is valuable for both industry professionals and researchers.
Welch and Wilkinson [27]	Samples from manufacturing firms in Shaanxi, Gansu and Xinjiang	To create a harmonious atmosphere, firms need to pay attention to power balance, conflicts resolution and cooperation.
Zolkiewski [18]	Three, in-depth case studies were undertaken in the UK National Health Service (NHS)	The cases utilized to exemplify the application of power within a network context offer a valuable catalyst for advancing both theoretical understanding and enhancing managerial proficiency. When viewed from a network perspective, it becomes evident that substantial effort has been dedicated to scrutinizing the bonds between actors, the connections between resources and the links between activities. In contrast, other influential network dynamics have not garnered as much scrutiny.
Helper [28]	U.S. and Japanese Auto Industries	Companies holding significant power in their end-product market can leverage that power to reshape the dynamics of their input markets. In a scenario where suppliers possess substantial bargaining power within a voice system, the company might discover it advantageous to transition to an exit system, enabling them to capture the benefits of their favorable position in the final product market.

Table 1.
Some industrial case studies with relevance of power issues.

their sales from Audi. The objective of this policy is to avoid an overly dependent relationship and to reduce risk. Following this policy, if one of the system integrators becomes too strong, Audi reduces its dependency by promoting a second-tier supplier to the first tier to fill the gap in supply. Using an opposite strategy, Suzuki usually has one supplier for each part. However, to maintain its bargaining power, Suzuki strives to keep the dependency of its suppliers through the volume of sales.

In the aeronautic supply chain described by Marcotte et al. [31], strategies are clearly influenced by implicit power relationships. This study was undertaken in the south-west of France, through interviews with supply chain managers, responsible for coordination aspects in companies of various tiers. Customers behave as dominators during the negotiation process because they know that they represent a significant cash flow for their suppliers. However, some suppliers, i.e., small and medium enterprises (SMEs) that produce high-value subassemblies and that also participate in other supply chains, do not analyze the power situation as the customers do. They argue that power stems for their much-needed mastered niche technologies. Such companies create difficult relationships in the supply chain hampering trust among partners and threatening the performance.

If these practical examples emphasize the importance of power in supply chain practices, they also illustrate the lack of a structured framework to assess the relative power of actors and its consequences in supply chain performance. In particular, the aeronautic example demonstrates how different assessments of a power situation can weaken the relationship between two actors and supply chain performance. Thus, it is important for companies to be aware of power, its properties and characteristics, to better analyze their collaborative situations.

Based on these aforementioned studies and those summarized in **Table 1**, we can conclude that the researchers and practitioners do identify power as a contributor to the substantial and stable operation of a partnership while it appears that further research is needed to actually address these problems [32]. The research presented in this chapter is motivated by these management situations in which companies find themselves, and the obtained results primarily concern modeling power and addressing the gap between the abstract concept of power and its application.

The objective of this chapter is to examine various aspects of power and the interactive process. To do so, we clarify and analyze the concept of power and its application. We will formally specify some properties of power more in terms of practical application than the philosophy of power dynamics. The formalization of these properties helps not only interested analysts and observers but also the actors in a particular power relationship to determine relevant collaborative strategies and tactics. This is due to this chapter extending beyond highlighting sources of power but also its instruments. In short, the primary contributions of the chapter are the following:

1. *Characteristics and properties of a power relationship*: We discuss some formal results modeling the concept of power and some of its characteristics formulated by algebraic properties. To the best of our knowledge, these efforts are the first attempts within the scientific literature to suggest a robust foundation of power theory formulated in an algebraic manner.

2. *Types of power*: According to the large state-of-the-art literature review we have done and the results on the characteristics and properties of power, we suggest a power taxonomy that clarifies the main differences between the classes of power that appear during a negotiation process among two actors or parties. This negotiation process can be analyzed as a power evolution process.

3. *Power evolution process*: The contribution of this study is to model the negotiation process from the point of view of power types evolution. We present here the stages of exercise of power in a negotiation between a company A and one of its customers (or supplier) B.

This chapter is organized as follows. Section 2 provides a literature review of power. The third section discusses the research methodology. Sections 4–6 present the findings: characteristics and properties of a power relationship, power types and the power evolution process. In Section 7 we illustrate the results of interviews of experts in a real supply chain between Morocco and Spain. Finally, managerial insights, conclusions and perspectives are presented.

2. Literature and related works

2.1 What is "power"?

This review provides a synthesis of various contributions on the topic of power. The concept of power is pervasive across multiple domains, extending beyond its role in social science research to also encompass areas such as industrial management, business strategy and marketing.

Koning et al. [33] argue that the concept of power is often considered as one of the more perplexing and elusive ideas, despite its frequent usage. This is to say, defining the concept of power in an uncontroversial manner has been proven to be difficult. When one party exerts control over another, it looks for achieving some desired outcomes. While power is commonly defined as one party's capacity to influence another, control should be seen as a result of the power exertion process, which may or may not yield success [34]. This is the consistent distinction between the ability to influence and the achieved influence [35].

Power is defined in various ways within different fields. In supply chain collaboration, two or more firms working jointly to plan and execute supply chain operations [2, 36], power is regarded as the ability of a firm to affect another's decision making and/or behavior [37]. Most definitions of power within studies on supply chains marketing channels are based on the definition of El-Ansary and Stern [38], who define power as "the ability of a channel member to control the decision variables in the marketing strategy of another member in a given channel at a different level of distribution". Power in supply chains is also defined as "the ability of a firm to own and control critical assets in markets and supply chains that enable it to sustain its ability to appropriate and accumulate value for itself by constantly leveraging its customers, competitors and suppliers". One of the most prominent studies on power was carried out by French and Raven in 1959. They identified five bases and forms of power that an actor (O) can exert over another actor (P). These different forms of power are defined below.

1. *Reward* power is defined as power whose basis is the ability to reward. The strength of the reward power of O over P (or O/P in short) increases with the magnitude of the rewards.

2. *Coercive* power of O/P stems from the expectation of P punishment (by O) if it fails to conform to the influence attempt.

3. *Legitimate* power of O/P is generated from the acceptance by P of the legitimate right of O to influence P.

4. *Expert* power: The strength of the expert power of O/P varies with the extent of the knowledge or perception that P attributes to O within a given area.

5. *Referent* power of O/P derives from P's respect for O and their desire to identify with or emulate him.

In the supply chain network context, coercive power [39] reflects the fear of a network member being put under more or less high pressure if it fails to comply with the requirements of the focal company [40]. The effectiveness of reward power hinges on the power holder's capacity to provide incentives to others [23]. When a focal company possesses resources that hold value for other participants in the network, it can compel these network actors to act in a manner that aligns with its objectives. Expert power emanates from an individual's or entity's proficiency or specialized knowledge in a specific subject. In the context of a supply chain network, a central company can attain expert power if the network participants believe it possesses unique technology, expertise or know-how that they deem valuable [41]. Legitimate power arises from a rightful authority to exert influence and an associated obligation for others to acknowledge and accept this influence. In this context, network members as having the legitimate right to make certain decisions [42] acknowledge a central actor. Referent power relies on the capacity to be appealing or likable to others, and it is contingent on the charisma and interpersonal skills of the power holder [43]. In the supply chain context, this power is observed when network actors want to join a network [23].

In the resource dependency theory [44], companies perceive that they are interdependent when they realize that they cannot control all necessary conditions to achieve desired outcomes. Therefore, power is based on the control of resources that are considered strategic within the organization [45]. The partners' dependency often leads to power imbalance and an advantage of one party that may limit the autonomy and constrain the behavior of the other party [37, 46, 47].

Martin [48] discusses evaluation techniques for power, stating that a major reason for the lack of research concerning power is the difficulty of measurement. In fact, it seems unrealistic to suggest a metric measurement for the power of a business. This is because power is an amalgam of complex social, economic and even psychological factors [15]. Power is a subjective phenomenon predicated on one's belief or expectation of how another actor will create an impact on oneself [49]. There were some attempts to assess the relative power by comparing two or more actors together, (see Cho and Chu [49], Zolghadri et al. [41]). In other words, the final conclusion that can be made regarding an actor involved in a relationship is either *stronger than* or *weaker than*. This comparison can be expanded to a larger scale.

According to all these aforementioned descriptions of power, in this context, power is defined as the ability to influence and mold the decision-making and actions of another party.

2.2 Use of power in supply chains

Koning et al. [33] agree that power differences greatly affect negotiation outcomes. He demonstrated that bargainers reached higher outcomes when they had more control over the decisions of the other party. The concept of power became popular in engineering industries when Porter [50] presented its five forces model. This model analyzes the structure of an industry and shapes the nature of competitive interactions within an industry. The supply chain context offers new perspectives to the study of power because of the presence of, for example, multilateral relationships, diversity of links and number of power inducers. The use of power can influence various aspects of the supply chain, such as buyer-supplier relationships, conflicts, actor performance, supply chain integration, etc. Coming back to the five types of power defined by French and Raven, some studies have shown that their usage did not reach to the same results in different situations or even countries. For instance, Yeung et al. [51] argued that the use of coercive power might have a detrimental impact on cooperation in Western cultures but can enhance supplier integration. In a general sense, a company can employ coercive power to strengthen internal coordination among buyers, which may subsequently foster supplier integration within a trust-based relationship. The research of Maloni and Benton [40] provides initial evidence that the exploitation of the supply chain by the power partner may lead to dissension and under performance thus hurting the power holder. Likewise, a judicious use of power may serve to benefit the power holder. Power is therefore not necessarily opposed to cooperation and should not always be observed in a negative way for a weaker party; see Peter Dapiran [52]. In this school of thought, several authors argue that power imbalances do not necessarily hinder the establishment of close and effective relationships, and they provide evidence that suppliers can attain improved economic performance and enhance their capabilities through collaborative relationships with influential retailers [53].

3. Research approach

The research we dealt with was exploratory. In fact, even if previous studies do exist, they rarely looked for formalizing the power properties as we desired to develop. Our goal was to find a set of properties and power patterns that could become the basis for further research. The techniques we used here include an extensive review of state-of-the-art academic and industrial studies and theoretical papers. Based on this extensive state-of-the-art review, we proposed a set of power properties and concepts necessary to build a power analysis framework that consists of describing and characterizing the power in an interactive process between two actors A and B (companies in a supply chain). Finally, an in-depth case study was performed to show the applicability of the framework. This case study was based on interviews as the main source of data collection, and it gave authors the opportunity of highly detailed discussions, in which the questions "why" and "how" were asked, and the most fundamental reasons and backgrounds behind the power patterns were explained by decision-makers, see [54]. The supply chain we focused on consists of an SME located in the center of Morocco, which is the focal company of the study, a software editor (also an SME located in SPAIN), a supplier (a very small business located in Morocco) and a customer (also an SME located in the north of Morocco). In these collaborative relationships, the focal company and its partners work together to coordinate the implementation of the information system for a large customer. We interviewed high-position persons (general directors, business consultants, technical directors, etc.) in each of these companies.

Given the approach of this study, the questions concerned the roles and capabilities of firms, their power positions and the nature of these relationships with other actors, sources and properties of power. The interviewees were asked to reveal some of their experiences in which the power of the customer or supplier lies behind the negotiations. No pre-formulation of concepts was suggested to the interviewees during these interviews to avoid any influence on their formulation of answers. Interviews remained semi-structured until the end. After completing each interview, we transcribed the answers to each question in a spreadsheet. We then analyzed the interview results and synthesized a series of findings and accompanying recommendations. The details about these interviews will be provided in Section 7.

4. Characteristics of power relationships

4.1 Characterization and a formal definition of power

After scrutinizing various aspects of power, it can be concluded that power typically pertains to the capacity or potential to influence or regulate the intentions, decisions or actions of others in order to further its own goals or interests. The various definitions across different sciences seem to use similar terms, apply the same common sense and reflect both the ability to control or influence with a difference of context in which they are applied. So, we formalize our understanding of power by the following definition: *Power is a property of a relationship established between two parties A and B [55]. Power of A over B, noted $P(\frac{A}{B})$, is A's conscious ability [56], derived from the possession of a source of power [54]*. This ability can be used or not used to control or influence B's behavior and decisions, even against its resistance. Possessing power does not automatically imply successful power exertion, which is nothing more than control. The control is therefore the successful application of power exertion. We can characterize a relationship generating power by.

1. *The sources of power* [57]: A has the ability to control or influence B because A possesses a source of power. Sources of power are the assets that form the basis of one partner's control over another. Let us denote $\boldsymbol{\sigma}(\boldsymbol{A})$ as the set of sources that A possesses. A's sources of power may encompass factors like A's knowledge, skills, scarce resources, expertise, and other attributes that are needed by B within a particular context. The power of any actor in a relationship is determined by the sources of power that the actor possesses at any given moment [38]. Ireland and Webb [58] cited some sources of power in supply chains: the number of major customers of a supplier's component, a supplier's market share of a given component, the number of suppliers from which a buyer purchases a particular component, the number of potential suppliers for a given component or the amount of revenue a supplier generates from a single buyer.

2. *The interactive context*: The interactive context between A and B defines the context of the relationship between them. By convention, if there are several relationships between A and B, we distinguish them by an index i (*i.e* $P_i(\frac{A}{B})$), for example two companies working together in two different supply chains.

3. *The resistance of B* [59]: Denoted $\boldsymbol{R}(\boldsymbol{B})$, the resistance of B against the power exercised by A is its ability to absorb the power exercised by A without changing its behavior or decisions. It is a reciprocal power (*i.e.*, $P_i(\frac{B}{A})$). Anderson and Weitz [60] note that there is ample evidence that the weaker party becomes mistrustful and apprehensive about the stronger party's intentions. The weaker partner consequently may even engage in a pre-emptive strike against the more powerful partner to protect its knowledge assets [15].

4. *The instruments of power* [61, 62]: Denoted $\boldsymbol{Ins}(\boldsymbol{A})$, the instruments of power are defined as tools or means used by A to exert its power over B. To explore A's instrument of power is to examine its operation, decision making and media accompanying this power.

5. *The tactics of use* [63–65]: Denoted T(A), the tactics of use is a set of methods and sequences of actions associated with the power instruments. These methods are defined by A to exert its power and represent how A plans to achieve its goals. For example, gaining market share from established competitors is at the top of most small businesses' strategic agendas. A number of marketing tactics can help to achieve this goal: (i) reducing prices in the growth stage can encourage consumers to try a company's product, (ii) offering promotions, such as coupons

or generous loyalty programs, also can encourage customers to try a company's product. In these tactics, the company uses as an instrument, e.g., the price, coupons or loyalty programs.

Finally, a formal description of power can be given below such that the power of A over B in an established relationship is a function of the relationship, the sources of power of A, the tactics and instruments used by A and finally the resistance of B:

$$P_i\left(\frac{A}{B}\right) = f(i, \sigma(A), T(A), Ins(A), R(B)) \quad (1)$$

4.2 The properties of a power relationship

This section analyzes power as a generic concept to characterize certain of its main properties. A first list of these properties is provided hereafter. This list of properties demonstrates that the power relationship is not naturally, randomly or spontaneously established. There is logic to the process, which can be explained or modeled by algebraic properties. These properties help to better understand the mechanisms of power relationships.

1. *Power is relational:* This property means that *there is no power without relationships between parties*. Power appears only in a situation where two or more actors are connected, it only exists within a relationship. Rojot [66] agrees that an actor is not powerful or weak in absolute but relative to another party.

2. *Power could be transitive:* This means that if two power relationships $P_i\left(\frac{A}{B}\right)$ and $P_i\left(\frac{B}{C}\right)$ exist, then the transitivity applies:

$$\left[P_i\left(\frac{A}{B}\right) \wedge P_i\left(\frac{B}{C}\right)\right] \Rightarrow P_i\left(\frac{A}{C}\right). \quad (2)$$

   The application of the transitivity rule is limited by three necessary conditions: (i) each actor A and C knows the existence of the other, (ii) the same relationship exists between all of the three actors, and (iii) actors A and C should be aware of $P_i\left(\frac{A}{B}\right)$ and $P_i\left(\frac{B}{C}\right)$. These conditions show that power is not always transitive. It means that if two different power relations $P_i\left(\frac{A}{B}\right)$ and $P_j\left(\frac{B}{C}\right)$ exist, no transitive conclusion can be drawn about the relationship between A and C. An example of transitivity is "the boss of my boss is my boss". On the contrary, "my boss has no relationship with my child" and therefore the transitivity has no place between them. In this case, the relationship could not be transitive.

3. *Power is not continuous:* The *non-continuity* property means that if A has power over B in a given moment t_i, there is no certainty about this relation remains the same in another time t_j.

$$\exists t_j \neq t_i \qquad \left[P\left(\frac{A}{B}, t_i\right)\right] \neq \left[P\left(\frac{A}{B}, t_j\right)\right] \quad (3)$$

For instance, in a case study of the relationship between a supplier S_1 and a customer C, two situations may be found for $P\left(\frac{C}{S_1}\right)$:

- At one instance in time, C needs to buy products based on technologies developed exclusively by the supplier S_1. In this case, S_1 can be considered as stronger than C because of its dependency to S_1's technology.
- If another supplier S_2 develops similar or comparable technologies to S_1, thus becoming a competitor, C can then choose between S_1 and S_2. In this new case, as the customer has the choice between two suppliers, the power imbalance is more favorable to C than before.

4. *Non-generalization of power:* The generalization is a reasoning that claims to demonstrate a universal proposition from particular examples. From a logical point of view, this argument reduces to $\exists x, f(x) \Longrightarrow \forall x f(x)$. No generalization can be made based on the power relationship between two actors. The first interpretation of this property is that if the supplier A_k has power over its customer B (i.e., $P\left(\frac{A_j}{B}\right), \forall j$), we cannot deduce that any supplier A_j is also stronger than B (i.e., $P\left(\frac{C}{B}\right)$). The second interpretation of this property is that is A has power over its customer B for a given relation i, we cannot deduce that the A has power over B for all relations.

$$P_i\left(\frac{A_k}{B}\right), A_k \in A = (A_1, \dots, A_m) \nRightarrow \left[P_i\left(\frac{A_j}{B}\right), \quad \forall i, j\right] \tag{4}$$

$$P_i\left(\frac{A}{B}\right) \nRightarrow \left[P_i\left(\frac{A}{B}\right), \quad \forall i\right] \tag{5}$$

5. *Power means reciprocity*: It means that no power relation can be considered as strictly unidirectional. This is to say,

$$P_i\left(\frac{A}{B}\right) \Rightarrow \left[P_i\left(\frac{A}{B}\right) \wedge P_i\left(\frac{B}{A}\right)\right] \tag{6}$$

The concept of reciprocity in power relationships brings up the question of assessment of $P_i\left(\frac{A}{B}\right)$ and $P_i\left(\frac{B}{A}\right)$. When there is inequality, the power model is referred to as a dominance model [47], *power imbalance* [36, 67] or *power asymmetry* [68, 69]. Power asymmetry pertains to the extent to which one firm possesses significantly more or significantly less power than another [69]. It is a scenario in which an actor is positioned to enforce its preferences and perspectives. In other words,

- If $P_i\left(\frac{A}{B}\right) > P_i\left(\frac{B}{A}\right)$ Then, A dominates B.
- If $P_i\left(\frac{A}{B}\right) < P_i\left(\frac{B}{A}\right)$ Then B dominates A.
- If $P_i\left(\frac{A}{B}\right) = P_i\left(\frac{B}{A}\right)$ Then we say that there is equality of power between A and B.

The power balance or imbalance is judged, more or less easily, by an observer and even by each of the actors. Each of them could conclude whether A is stronger or weaker than B or that they have equal power. Each actor or the observer can evaluate the power imbalance using subjective and objective evaluation processes.

5. Various types of power

Power has several types. These types are situations in which power achieved certain conditions. These conditions may be characteristics, decisions or actions of an actor. This section presents the different types of power that are perceived during the interactions between actors (see **Figure 1**).

5.1 Potential power

Potential power is "the underlying capacity of negotiators to obtain benefits from their agreement" [70]. This definition shows that potential power is an ability not necessarily used. Potential power could be a source of influence even when it is not exercised. For example, the more dependent firm may comply with something it believes that the more powerful firm wants, even if this latter firm may never make the request [23]. The potential power of A over B is a function of the dependence of B upon A (B's dependence on A and thus A's potential power over B). Almost all research works agreed on the fact that the power is derived from dependence [45]. This assertion is based on the pioneering work about power in organizations conducted by Cook et al. [71] and Emerson [72]: "power resides implicitly in the other's dependency". In the theory of resource dependence, as presented by Pfeffer and Salancik [44], dependency arises from the exchange process and the need for organizations to obtain resources and engage in exchanges with their environment. Within a supply chain, an actor can gauge the value of the resources they provide and the dependencies that result from these exchanges. Resources are broadly defined to include people, assets, materials or capital used to achieve specific goals, whether they are human, intellectual or financial in nature. Stronger members within the supply chain may assess the dependencies of others, thereby gauging their potential power.

In cases of mutual dependency, it becomes more or less crucial for each party to have the ability to control or at least influence the behavior of the other. Simultaneously, these mutual dependencies imply that each party has the capacity to grant or deny, facilitate or obstruct the satisfaction of the other party's needs or desires.

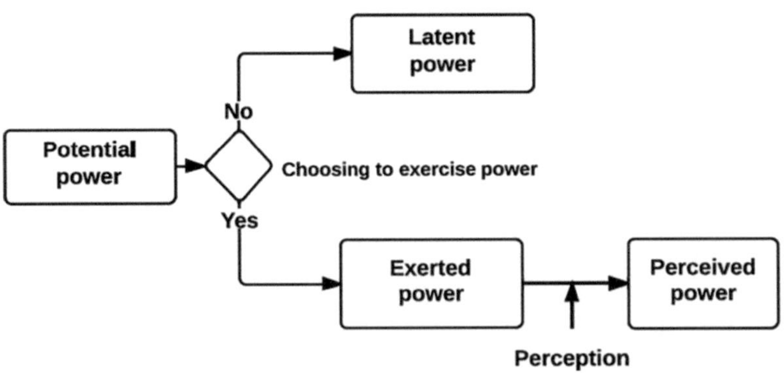

Figure 1.
Types of power.

5.2 Latent power

In certain instances, a party may be aware of its own strength but chooses not to leverage its power for personal advantage or to manipulate the actions of the weaker party. This is a case of latent power. It is a consequence of choosing to not use potential power. Some researchers [62, 73–75] have proposed that power exists in its potential, while others have argued that power is only evident in its utilization (an actor with power should demonstrate a greater propensity for action, irrespective of the social consequences of their actions). The differentiation between possessing power and actively employing it is of significance when investigating relationships within a supply chain.

For example, a customer can influence the decisions made by a company without exercising its power. This means that the company does account for potential (negative) reactions of its customer while making decisions. A does not do anything to influence B's behavior; the weaker party simply anticipates the needs and wishes of the more powerful one.

5.3 Exerted power

Exerted power, denoted $\tilde{P}$, is the conscious use of potential power using adopted protocols and instruments. Exerted power is directly connected to the will of a party to use its potential power. This is to observe that having power does not necessarily imply its use, but to exercise power requires possessing potential power. The decision of whether to wield power hinges on a party's goals and its negotiation approach. Spangler [76] makes a distinction between positional and interest-based bargaining, emphasizing that real-world business negotiations likely involve a blend of these two strategies. Positional bargaining is characterized as an adversarial negotiation strategy, often employed by negotiators who view negotiations as a win-lose scenario. In contrast, interest-based bargaining is a collaborative strategy, with both parties aiming to collectively fulfil each other's needs, resulting in a win-win outcome.

If one partner is stronger than the other and has the propensity to exert its power then the parties will be engaged in a positional negotiation, and the stronger party will obtain a significant advantage, while the weaker will have small benefit from collaboration. In this case, the weaker actor has no or low capability to push negotiations from a positional negotiation towards an interest-based one if the stronger partner does not wish it. The only possible solution for the weaker company to transform both situations is to apply clever negotiation techniques. If one partner is stronger than the other, and it does not choose to exert its power (according to its business philosophy for instance), then the parties will be involved in an interest-based negotiation, and benefits for the weaker party can be guaranteed. Acting based on power might increase the tendency towards action even when power is not directly possessed, applicable, or relevant. Exerted power allows an individual to engage in actions that can promote successful completion of goals.

5.4 Perceived power

The exercise of power depends on its tactical use by A to influence B's decisions and behavior. However, B may not understand this power exertion by A or it could misunderstand A's will. In this case, it is possible for B to not comply with A, and choose to act according to B's own will. Here again, a conflict could appear between the actors A and B, see Marcotte et al. [31]. Therefore, another critical factor in the process of application and use of power is the perceived power by the actor, subject to the process exertion process, which is known here as B. For Ford and Johnson [77], the perception of power is a step in a process in which power capabilities may or may not be used and, if used, have an uncertain probability of success. The process of perception is never objective; it is the interpretation that gives meaning to the power received. The success of the A's power exertion process is thus profoundly conditioned by perception of B.

6. Power evolution process

A power relation between two actors is created after the establishment of a relationship between them. The relation establishment and the consequent power relation are modeled through the evolution process shown in **Figure 2**. We suppose that a rational actor prepares its negotiations with another one. Each party tries to rationally use its advantages to convince or persuade the other party [5]. In our case, A actively uses its power to influence or control B. This process involves several steps depicted in **Figure 2**: column A corresponds to all actions concerning the actor A, the column A and B correspond to all actions that require intervention and the existence of two actors and the column B corresponds to all actions specific to B.

This process, namely, the "power evolution process", describes various scenarios in which a party develops these characteristics and advantages in a given context to potential power and how this power is transformed to other types of power (exercised, latent and perceived). This process can be divided into four stages described in the following.

1. *Defining potential power*: At this stage, A looks to determine its position regarding B through the evaluation of its sources of power that could give it advantages over B.

2. *The situational analysis*: The environment in which A is located plays an important role in strategic preparation. Once A determines its position and its potential power, the second step in this process is to analyze the environment in terms of rules, knowledge, beliefs, goals and any constraints that may limit the actions of A. This is a necessary step before making decisions if one expects to have good results [78]. The purpose of this situation analysis is to determine the characteristics of the internal and external environment that directly affect A's strategic plan. It also allows A to be aware of these characteristics in its decision-making tasks. A's analysis effort focuses on the generation of answers to a series of strategic questions and then using these responses to form an image of the strategic situation [79]. At the end of this analysis, the party would then be able to make a decision on the exercise of power. If the party decides to keep this power, the potential power is transformed to latent power. On the contrary, if the party decides to exercise its power to attain these objectives, it will move onto the analysis of tactics.

3. *Analysis of tactics and the exercise of power:* This step involves the transformation of potential power to exercised power. This phase models the way that actor tries to leverage existing capabilities and its potential power to improve its situation and to achieve its goals. To execute its power, A makes an analysis and chooses

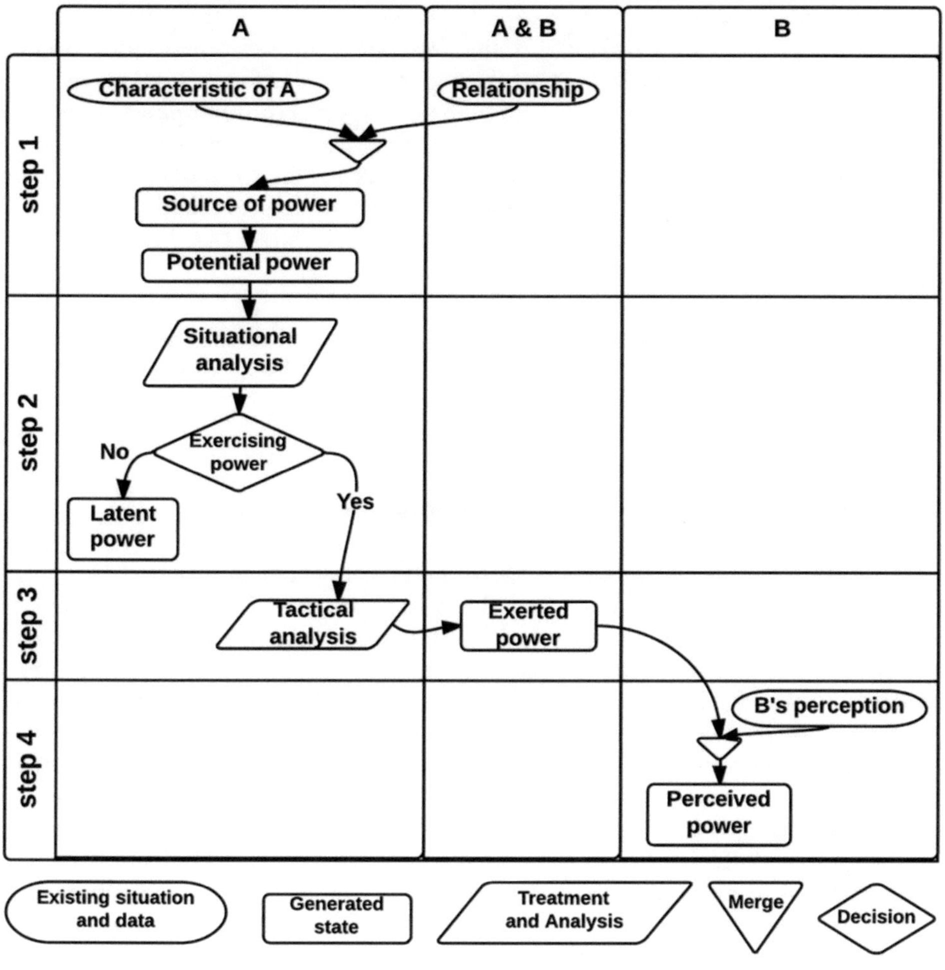

Figure 2.
Power evolution process.

the best tactics to convince B. The tactical use of power analysis includes the definition of a set of maneuvers representing how a party achieves a result. These tactics are developed by the actor (initiator of the negotiation) taking into account (1) the positions of two actors, (2) past relational negotiation history and (3) the available instruments. We can distinguish two types of tactics that the stronger actor can take to exercise its power:

- *Coercive tactics:* They are used with the aim of bringing the other actor to adopt a desired behavior. The goal of this type of tactic is to exert influence on its partners based on exercising force [61] with the intention of controlling negative outcomes relative to each other, either through punishment such as cutting the price or terminating the relationship, see Yeung et al. [51] for more details. The use of non-controllable coercive tactics may negatively influence the partnership increasing conflicts, reducing satisfaction and the weaker actors' willingness to cooperate [37].

- *Persuasion and convincing tactics*: Its goal is to convince the other actor to make concessions by trying to demonstrate that the action request is not only justifiable but important through the distribution of rewards or benefits (written contracts, lobbying of interests through power structures at a legislative level, negotiations and discussions, investments into the production, financial assistance to producers in the form of credit or leasing, trainings and educational activities). For example, the relationships established by Toyota with some promising suppliers (see Liker [80]) are examples of these persuasion exercises of power.

4. *The perception of exerted power*: Finally, the success or failure of power exertion depends directly on the perceived power by B. A implicitly or explicitly should show B its goals by the exercise of power.

7. Illustrative case: a high-tech supply chain in Morocco

The illustration of these results was performed through a deep supply chain analysis mainly by experts' interviews. The selected company, denoted Fc, is an SME firm located in the center of Morocco involved in information system engineering. Fc sells its expertise and software packages to administrations, companies from the oil industry, associations and sport clubs. For the purpose of simplification in the collaboration analysis, one specific project was selected. In this project, Fc coordinates the implementation of the information system for a customer. The customer is also an SME denoted hereafter Cu for customer. Fc has to implement, manage and control the whole information system of Cu and to do so, Fc uses the software package sold by a software editor Ed, which is again an SME. Finally, Fc buys the required networking equipment from Su. Fc and Su already have an established relationship from a series of past successful collaborations. These companies are of various sizes and could have potential power in some situations, and they look to exert it whenever possible in an opportunistic way. Fc, Su and Cu are located in different areas in Morocco while Ed is in Spain (see **Figure 3**). We interviewed high position persons (general directors, business consultants and technical directors) in each of these companies. Our goal was to find out.

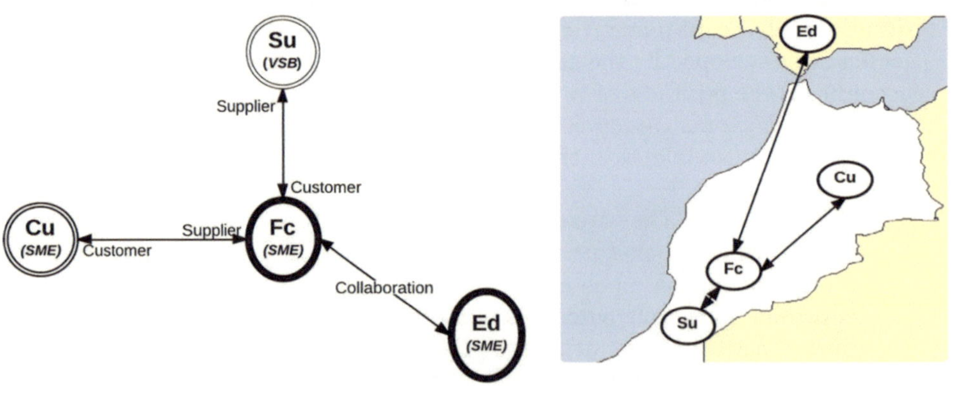

Figure 3.
Fc's network.

1. Whether power plays a crucial role, in an apparent or hidden manner, in their negotiations or not?

2. What were the main power sources, gains and losses of parties, instruments?

3. What are the power properties that can be illustrated through the use-cases?

After obtaining their consent to participate, the four companies were first asked to evaluate a draft questionnaire. The consolidated questionnaire was used as a means to guide the interviews. Four semi-structured interviews were conducted containing open-ended and closed-ended questions. Interviewees revealed their point of view regarding their dependency and power relationship.

Focusing on the high-tech sector was a conscious decision to target an area where power relations are highly dynamic leading companies to frequently analyze their own relative position in the supply chain. Coincidentally, this is linked to the concept of clock speed developed in Fine [81]. In short, the reasons for focusing on the high-tech sector are:

Due to the rapid evolution of software and hardware technologies, also known as clock speed, the high-tech sector is characterized by highly dynamic behavior, which also extends to power relationships within the sector's supply chains. This is to say, that power imbalances do not hold steady for long periods of time. The sources of power evolve, and the owners of these sources must carefully analyze their network on a frequent basis. Furthermore, companies in this sector have to address power issues much more frequently than other sectors.

The sources of power could even offer strong power positions to small firms, primarily in terms of mastered technologies.

Customer satisfaction is measurable with little latency due to the service rate. This yields a stable collaboration establishment equally challenging for all the supply chain actors.

The use or abuse of power could result in quick changes to the supply chain structure, i.e., choosing other partners from a large pool of potential ones.

7.1 Main results and discussion

The interviews we made allowed us to identify the interdependences and power imbalances between companies. Here, we will browse the power relationship between *Fc* and each of the three actors *Cu*, *Su* and *Ed*. It should be noted that asking actors to express their own power is an important bias. Indeed, the notion of power is not a neutral concept and seems somehow to force actors to emit a judgment.

7.1.1 Facts and sources of power

Browsing the companies, the following observations can be made:

- *Su and Fc*: *Su* sales mainly networking equipment and offers *Fc* payment facility reaching 90 days. No other customer is allowed to do so. For *Su*"*Fc is an important customer. We know that Fc works on special projects that demand a large amount of materials: server, pc, printers, etc. We must satisfy Fc*". (Extracted from the interview of the *Su*'s CEO).

- *Cu and Fc*: *Cu* seeks to implement a relevant information system. They chose *Fc* as the supplier of their information system based on its extensive expertise, quality of service and impressive number of customers (40) who used the same software that *Cu* needs. From the *Fc*'s point of view, *Cu* is also an important strategic door opener: *"We have many customers in the south of the country. Cu will give us a wonderful opportunity to expand our market towards the north"* (Extracted from the interview of the *Fc*'s CEO).

- *Ed and Fc*: *Ed* is a multinational company known by its capacity and expertise in the area of integration and implementation of information systems. *Fc* has developed a partnership agreement with *Ed* to be able to offer combined software and services. This forms a high level of synergy between the two companies. Through this program, *Ed* provides technical supports to *Fc* and allows *Fc* to gain customers' satisfaction. In tandem, *Fc* gives *Ed* an opportunity to expand its market into northern Africa.

Table 2 shows the main sources of power identified and validated by each of the companies.

7.1.2 Properties of power in Fc*'s network*

In this section, the transitivity, non-continuity, non-generalization and reciprocity of power relationships can be illustrated through the analysis of the *Fc*'s supply chain supported by some extraction of the managers' interviews.

1. *Transitivity property*: *Fc* has noticed that *Su* is often late to deliver orders. After analysing the situation, *Fc* understood that orders are often late because the equipment wholesaler does not give the *Su* orders high priority. According to *Su*'s CEO, this is not due to strategy. In fact *"we do not seek to exploit Fc, but you should know that I also suffer a pressure of my wholesaler"*. The equipment wholesaler, denoted *W*, uses its power over *Su* knowing that its equipment price is such that *Su* does not desire to change for any other wholesaler in the area. In its turn, *Su* imposes its delivery conditions to *Fc*, which is impacted by this low performance. *Fc* depends strongly on *Su* and does not have any means to change the power imbalance because of the payment facility offered by *Su*. Therefore, there is a transitivity property linking the wholesaler (even if it is outside of the considered supply chain) to *Fc* via *Su*.

Actor	Source of power
Fc	Expertise, quality of service, number of customers, geographic position and its ability to purchase large quantities of equipment.
Cu	Brand image and geographic position.
Ed	Technical and strategic expertise, multinational company, expertise in the areas of integration and implementation of information systems.
Su	Payment facilitating.

Table 2.
Source of power in Fc's network.

$$\left[P\left(\frac{W}{Su}\right) \wedge P\left(\frac{Su}{Fc}\right)\right] \Rightarrow P\left(\frac{W}{Fc}\right). \tag{7}$$

2. *Non-continuity property*: According to the interviews of the *Fc* top managers, their relationship with *Cu* is quite balanced. However, from the point of view of *Cu*, the situation is different. They chose *Fc* through a long selection process with several competitors. At that moment, let us denote it *initial*, *Cu* believed itself much stronger than *Fc*. However, once *Cu*'s decision of using the *Fc*'s software made, they became aware of their need for training and maintenance. In this new situation, referred to as *final*, *Cu* became highly dependent on *Fc*, meaning the power balance was also changed (*Fc* became stronger than *Cu*), i.e., the two companies have had power over the other at different times of collaboration

$$\left[P\left(\frac{Cu}{Fc}, initial\right)\right] \neq \left[P\left(\frac{Cu}{Fc}, final\right)\right] \tag{8}$$

3. *Non-generalization:* From the illustration of the past property (non-continuity) it can be concluded that the evolution of power relation between *Cu* and *Fc* does not allow any generalization of a power pattern. A favorable situation to *Cu* can be transformed into a less favorable one by including a new source of power by *Fc*. Therefore, it is not possible to say that *Cu* is always in a dominant position regarding *Fc*.

4. *Reciprocity Property and power imbalance*: For the CEO of *Fc* *"the relationship with Su is based on commitment and trust"*. *Fc* is satisfied with its cooperation with *Su* (the equipment seller). Nevertheless, regarding the power imbalance, their level of dependence is different. For *Fc*, *"we present only 10% of the turnover of Su, while Su provides us 79% of our purchases"*. Therefore, both companies are in a bidirectional power relation, i.e., they are interdependent.

$$P\left(\frac{Su}{Fc}\right) \Rightarrow \left[P\left(\frac{Fc}{Su}\right) \wedge P\left(\frac{Su}{Fc}\right)\right] \tag{9}$$

7.2 The use of power in the Fc's supply chain or power evolution process

7.2.1 Exerted power

The analysis of the power evolution process within the supply chain shows that almost all of the potential powers are exerted in the collaboration relations. However, it will be observed that the situations are different between the couples *Fc-Su*, *Fc-Cu* and *Fc-Ed*.

1. *Fc-Su:* The power relationship established between *Fc* and *Su* is based on purchasing and bargaining strategies. *Fc* seeks to minimize the project's purchase cost while *Su* seeks to increase the selling price and the net benefit margin. For *Fc*

"Our supplier, *Su*, is aware of its power source. So sometimes they try to use their power to increase their margin during contract negotiations. This is especially the case when we need specific products. In this case, *Su* often tries to increase its profit margin". This means that in some situations *Su* uses its potential power, thus transforming it into exerted power. According to the CEO of *Fc*, *Su*'s power exertion is admitted and successful. No specific collaboration complication or frustration was noticed from this situation, which is accepted as a matter of fact (reward and referent power, see French and Raven [61]).

2. *Fc-Cu:* The evolution of power is more visible in the *Fc-Cu* relationship. These two companies have power over the other over different periods of collaboration, and each one seeks to gain the maximum benefits from the other. *Cu* requires of *Fc* paid penalties if it does not respect the terms of the contract (coercive power in French and Raven [61]), and on the other side, *Fc* has the ability to instruct *Cu* to buy other services and to accept another contract for training and maintenance. The analysis of the power patterns can be subdivided into two phases:

   - *Pre-contract establishment phase*: In this phase, the customer *Cu* has the main source of power, which is to choose among a series of similar product providers. Often, the suppliers of such solutions are high-tech SMEs needing contracts to launch their activities on a bigger scale. The potential suppliers, such as *Fc*, have to have enough sources of power to be more attractive than others. The number of established customers, the reputation, for their technologies is among some of the *Fc*'s sources of power. *Cu* has a strong potential power, transformed continuously into exerted power, against all of the potential suppliers. According to the situation described by *Cu*'s managers, they used a precise and long partner selection process. The description of hard discussions between *Cu* and *Fc* at the very end of negotiation process gives proof of power exertion by the customer. The situation is accepted as a normal process by *Fc* because of the non-continuity of the power and the new situation during the second phase.

   - *After-contract establishment phase*: Once the contract was signed, *Fc* obtained a much more favorable situation because of the creation of the dependency to its expertise and implemented technology.

All in all, both companies use their potential power before and after the contract signature. Their potential power is fully transformed and the instruments of power (formal contract conditions from one side and training needs from the other) are applied. No one seems to use the potential power as a latent one.

3. *Fc-Ed:* According to the terms of this partnership between *Fc* and *Ed*, *Fc* must pay annually an amount of money to *Ed* to continue this collaboration. *Ed* offers *Fc* reductions based on the number of new customers. Therefore, *Ed* uses its power by coercive (annual costs) and persuasion and convincing tactics (discounts and providing technical support to *Fc*).

7.2.2 Analysis of power exertion situations and management drivers for Fc

In the relationship *Fc-Su*, the power exertion does not create a negative reaction of *Fc*. In the *Fc-Cu* relation, the challenge is admitted as the "normal" business game between two collaborating actors. Each of them looks for more benefit while seeking long-term collaboration. From our observer's point of view, it might be observed that *Fc* is more dependent to *Su* than the other way. Therefore, *Su* has a stronger potential power than *Fc*; the collaboration is unfavorable to *Fc* forcing it into concessions.

We noticed a contradiction between the announced terms of collaboration for *Ed* as an international partner looking for new market in Morocco and its acts. In fact, it can be observed that *Ed* uses a coercive instrument to force *Fc* in its collaboration. This can be analyzed as a tough or even aggressive management approach, and further analysis seems to be necessary to relieve *Fc* from the exerted pressure.

The analysis of exerted power and sources of power of all the concerned actors, from the point of view of *Fc*, provided in this part can allow us to conclude upon some key management drivers. In fact, it seems necessary for *Fc* to use the results of this analysis to build a bottom-up business and operations strategy. As the power exertion in the context of this supply chain is usual, if *Fc* looks for creating sound relationships with others, it would be worth pursuing analysis of its own sources of power and also those of other these companies. This set of results could allow *Fc* to become more aware of its weaknesses and set up measures to balance the relations.

The management conclusions for *Fc* can then be summarized in what follows:

1. *Fc* has to look for a more balanced relation with *Su*. Even in the short term, the source of power of *Su* (i.e., the payment facility) allows *Fc* to have a better cash flow to dealing with liquidity problems: the late and long delivery delays would harm its business because of possible better conditions offered by its competitors to customers such as *Cu*.

2. *Fc* has to have long term collaboration with *Ed*. Again, due to the fragile economic situation of SMEs in the high-tech sector, *Fc* does accept conditions (penalties) that reduce the benefits. Nevertheless, in long term, *Fc* should identify competitors to *Ed* or any other strategies, such as developing in-house software, which would help to reduce its dependency towards *Ed*.

3. *Fc* and *Cu* have a balanced situation in the long run. Through an analysis of the whole business process, *Fc* can be aware of its concessions during the pre-contract phase while it could have a better position in the post-contract phase. The power analysis therefore shows that no mistakes can be made during the first round of negotiations (i.e., *Fc* does have to accept certain tradeoffs) because of better conditions in the second phase.

8. Managerial insights

In general terms, for a given collaboration situation in supply chains, a power-based analysis methodology is designed, where the steps are the following:

1. Initialization: Identify the actors building the supply chain viewed from the point of view of a focal company. For each of these actors, search for identification of sources of power, interaction contexts, instruments and types of power.

2. Analysis: Analyze the power properties to determine explicit and hidden power relations.

3. Strategies design: Define strategies to address reducing dependencies by designing leverage and counter-power strategies. These strategies are based on the focal company's sources of power and its need for collaboration.

4. Tactics design: Define tactics for the application of the defined strategies.

The strategies can be therefore of two classes:

- If the company is stronger than the partner, several leverage strategies are possible. For one, power can be exerted using various instruments to make the situation even better. This short-term strategy could be transformed into the most dangerous tool to cut off all collaborations. Another modern leverage strategy available is to use as a leverage tool by the stronger partner to help the weaker one reach higher quality and an improved management system, as demonstrated by the case of the Toyota Production System and one of the 14 rules of the Toyota Way (see Liker [80]).

- If the company seems to be weaker than the partner, several issues are possible: (i) accepting the deal while looking to reduce the consequences of a relationship imbalance by using clever negotiation techniques, (ii) reducing, to the extent possible, the dependency by using alternative technologies, persons, tools, methods, locations, etc. These are the counter-power strategies that have to be designed. Being able to do so means being able to accurately identify power patterns, sources, gains, loses and power exertion instruments.

9. Conclusions and perspectives

Analyzing power dynamics between partners is a relevant methodological approach for comprehending the behavior of each partner and its impact on the overall partnership strategy. This paper describes the characteristics of power and the process that leads to its possible use. This chapter gives a definition of power in the context of a supply network. We presented a set of elements and properties necessary for understanding power relationships. Finally, a taxonomy of various types of power has been established and linked to the process of potential use of its exertion.

Comprehending power dynamics holds direct significance for managers and practitioners, as it offers a more profound insight into the sources of power within organizations. In a broader context, the insights derived from power relations should be integrated into an analytical approach that offers valuable perspectives, similar to a SWOT analysis [82]. For instance, through the definition of reciprocal power, it can be understood that even a weaker partner has some type of power over the stronger one. A possible exertion of power from the weaker side of a power relationship, if the stronger partner looks to apply pressure upon the weaker one, is to cut the relationship by looking for other partnerships. Even if this is not always possible due to the business context, it remains a powerful way to eradicate the frustrations created in such imbalanced links. Obviously, using power does not always imply that coercive actions have to be taken. Instead, knowing that power includes persuasion and convincing tactics might lead to a change in behavior enhancing cooperation. This is particularly valuable because chain management is the alignment of actions in accordance to one's interests.

To be able to analyze and manage the relationships in a supply chain, managers need to be aware of characteristics, properties and types of power. From the perspective of supply managers, this is an issue worth studying given the need to understand the role of power as an influencing factor in the conception, development and maintenance of collaborative relationships. Moreover, by enhancing the managers' knowledge about power, such studies help companies in their supply strategy development. If managers are able to acknowledge power behind relations, they are able to classify their relationships and develop leverage or counter-power strategies on how to handle and manage different types of relations. In such relationship developments, managers will be aware of factors (sources of power) that should be highlighted and considered as the most crucial ones, if their own power position requires development.

The results of the study are also applicable in the process of supplier selection. If supply managers are aware of power in each relationship, they can use this knowledge when selecting supplier among many alternatives. Understanding the relative power towards a supplier is particularly important in cases when the supplier is new and the interaction is planned over an extended period of time. With established suppliers, power might become a means for leverage in contract or price negotiations [14].

This article looks for answering some fundamental questions concerning future research on power and its impact on the design and implementation of a product in a network of partners: how to assess and characterize imprecise fuzzy power, and in more concrete terms, how to model a mathematical relationship of power and how to deduce actors that govern the network. Through the application of a power-based analysis methodology applied to a real case, it is possible to clearly identify not only certain basic power properties (such as reciprocity, transitivity and non-continuity) among actors but also to find out how the sources of power could affect the relative position of one partner regarding the other [71]. The findings support the insights of Kahkonen and Virolainen [54], Cendon and Jarvenpaa [19], He et al. [15] and Zolghadri et al. [14] who called for research on power in networks.

Further developments should be performed to consolidate the power theory, as formulated in this study. Facts, interviews and industrial studies would help to make these primary concepts more robust on a considerably larger scale.

Author details

Driss Essabbar[1*], Maria Zrikem[2] and Marc Zolghadri[3]

1 LAMGIP EMSI (Moroccan School of Engineering Sciences) Marrakech, Morocco

2 TIM Laboratory-ENSA, Marrakech, Morocco

3 Phenics-LISMMA Laboratory EA2336 Supmeca, Paris, France

*Address all correspondence to: driss.essabar@gmail.com

References

[1] Arzu A, Erman E. Supply chain performance measurement: A literature review. International Journal of Production Research. 2010;**48**(17): 5137-5155

[2] Cao M, Zhang Q. Supply chain collaboration: Impact on collaborative advantage and firm performance. Journal of Operations Management. 2011;**29**(3):163-180

[3] Munksgaard KB, Stentoft J, Paulraj A. Value-based supply chain innovation. Operations Management Research. 2014; 7(3–4):50-62

[4] Su Q, Song Y, Dang J. The impact of supply chain relationship quality on cooperative strategy. Journal of Purchasing and Supply Management. 2008;**14**:263-272

[5] Chang CW, Chiang DM, Pai FY. Cooperative strategy in supply chain networks. Industrial Marketing Management. 2012;**41**(7):1114-1124

[6] De Faria P, Lima F, Santos R. Cooperation in innovation activities: The importance of partners. Research Policy. 2010;**39**(8):1082-1092

[7] Akintoye A, McIntosh G, Fitzgerald E. A survey of supply chain collaboration in the UK and management construction industry. European Journal of Purchasing and Supply Management. 2000;**6**(3):159-168

[8] Bahinipati B, Deshmukh S. Vertical collaboration in the semiconductor industry: A decision framework for supply chain relationships. Computers and Industrial Engineering. 2012;**62**(2): 504-526

[9] Michaelides R, Morton SC, Liu W. A framework for evaluating the benefits of collaborative technologies in engineering innovation networks. Production Planning Control. 2013;**24**(2–3):246-264

[10] Weitzel T, Beimborn D, König W. Coordination in networks: An economic equilibrium analysis. Information Systems and e-Business Management. 2003;**1**(2):189-211

[11] Zhang Y, Dilts D. System dynamics of supply chain network organization structure. Information Systems and e-Business Management. 2004;**2**(2–3): 187-206

[12] Tost LP, Gino F, Larrick RP. Power, competitiveness, and advice taking: Why the powerful don't listen. Organizational Behavior and Human Decision Processes. 2012;**117**(1): 53-65

[13] Fast NJ, Sivanathan N, Mayer ND, Galinsky AD. Power and overconfident decision-making. Organizational Behavior and Human Decision Processes. 2012;**117**(2):249-260

[14] Zolghadri M, Amrani A, Zouggar S, Girard P. Power as a high level criterion in partners selection process. International Journal of Computer Integrated Manufacturing. 2011a;**24**(4): 312-327

[15] He Q, Ghobadian A, Gallear D. Knowledge acquisition in supply chain partnerships: The role of power. International Journal of Production Economics. 2013;**141**(2):605-618

[16] Benton W, Maloni M. The influence of power driven buyer/seller relationships on supply chain satisfaction. Journal of Operations Management. 2005;**23**(1):1-22

[17] Leonidou LC, Talias MA, Leonidou CN. Exercised power as a driver of trust and commitment in cross-border industrial buyer seller relationships. Industrial Marketing Management. 2008;**37**(1):92-103

[18] Zolkiewski J. The complexity of power relationships within a network. In: Proceedings of the 17th Annual IMP Conference; Oslo Norway. 2001

[19] Cendon B, Jarvenpaa S. The development and exercise of power by leaders of support units in implementing information technology-based services. The Journal of Strategic Information Systems. 2001;**10**(2):121-158

[20] Porter ME. Choix stratégiques et concurrence. Paris: Editions Economica; 1982

[21] Trienekens JH, Beulens AJM. Views on inter-enterprise relationships. Production Planning Control. 2001;**12**(5):466-477

[22] Ming Y, Grabot B, Hou R. A typology of the situations of cooperation in supply chains. Computers Industrial Engineering. 2014;**67**(1):56-71

[23] Belaya V, Hanf JH. Using power in business relationships-old wine in new skins. Management Marketing. 2013;**8**(1):3-24

[24] Munksgaard KB, Clarke AH, Storvang P, Erichsen PG. Product development with multiple partners: Strategies and conflicts in networks. Industrial Marketing Management. 2012;**41**(3):438-447

[25] Wei J. Power relationships and schedule flexibility—Case studies on two supply chain modules. In: Proceedings of the 1st International Conference on Logistics, Informatics and Service Science. Vol. 2. 2011. pp. 58-63

[26] Zhuang G, Herndon N, Zhou N. Exercises of power in marketing channel dyads: Power advantage versus power disadvantage. The International Review of Retail, Distribution and Consumer Research. 2006;**16**(1):1-22

[27] Welch C, Wilkinson I. Network perspectives on interfirm conflict: Reassessing a critical case in international business. Journal of Business Research. 2005;**58**(2):205-213

[28] Helper S. Comparative supplier relations in the U.S. and Japanese auto industries: An exit/voice approach. Business and Economic History. 1990;**9**(2):153-162

[29] Shi D. A review of enterprise supply chain risk management. Journal of Systems Science and Systems Engineering. 2004;**13**(2):219-244

[30] Demeter K, Gelei A, Jenei I. The effect of strategy on supply chain configuration and management practices on the basis of two supply chains in the Hungarian automotive industry. International Journal of Production Economics. 2006;**104**(2):555-570

[31] Marcotte F, Grabot B, Costa Affonso R. Cooperation models for supply chain management. International Journal of Logistics Systems and Management. 2009;**2**(19):123-153

[32] Meehan J, Wright GH. Power priorities: A buyer-seller comparison of areas of influence. Journal of Purchasing and Supply Management. 2011;**17**(1): 32-41

[33] Koning L, Steinel W, van Beest I, van Dijk E. Power and deception in

ultimatum bargaining. Organizational Behavior and Human Decision Processes. 2011;**115**(1):35-42

[34] Handley SM, Benton WC. Mediated power and outsourcing relationships. Journal of Operations Management. 2012;**30**(3):253-267

[35] Mohr JJ, Fisher RJ, Nevin JR. Collaborative communication in interfirm relationships: Moderating effects of integration and control. Journal of Marketing. 1996; **60**(3):103

[36] Edirisinghe N, Bichescu B, Shi X. Equilibrium analysis of supply chain structures under power imbalance. European Journal of Operational Research. 2011;**214**(3):568-578

[37] Gaski JF. The theory of power and conflict in channels of distribution. The Journal of Marketing. 1984;**48**(1):9-29

[38] El-Ansary AI, Stern LW. Power measurement in the distribution channel. Journal of Marketing Research. 1972;**9**(1):47-52

[39] Zhao X, Huo B, Flynn BB, Yeung JHY. The impact of power and relationship commitment on the integration between manufacturers and customers in a supply chain. Journal of Operations Management. 2008;**26**(3): 368-388

[40] Maloni MJ, Benton W. Power influences in the supply chain. Journal of Business Logistics. 2000; **21**(1):42-73

[41] Zolghadri M, Eckert C, Zouggar S, Girard P. Power-based supplier selection in product development projects. Computers in Industry. 2011b;**62**(5): 487-500

[42] Munson CL, Rosenblatt MJ, Rosenblatt Z. The use and abuse of power in supply chains. Business Horizons. 1999;**42**(1):55-65

[43] Flynn BB, Zhao X, Huo B, Yeung JHY. We've got the power! How customer power affects supply chain relationships. Business Horizons. 2008; **51**(3):169-174

[44] Pfeffer J, Salancik GR. The External Control of Organizations: A Resource Dependence Perspective. New York: Harper & Row; 1978

[45] Zhuang G. The relationship between power and dependence in marketing channels. European Journal of Marketing. 2004;**38**:675-693

[46] Casciaro T, Piskorski MJ. Power imbalance, mutual dependence, and constraint absorption A closer look at resource dependence theory. Administrative Science Quarterly. 2005; **50**(2):167-199

[47] Essabbar D, Zolghadri M, Zrikem M, Ayadi A. Rule-based modelling of dependencies among activities in collaborative partnership. Journal of Design Research. 2014;**12**(1/2):32-51

[48] Martin R. Bargaining Power. Oxford: Clarendon; 1992

[49] Cho DS, Chu W. Determinants of bargaining power in OEM negotiations. Industrial Marketing Management. 1994; **23**(4):343-355

[50] Porter ME. How Competitive Forces Shape Strategy. Cambridge: Harvard Business Review; 1979

[51] Yeung JHY, Selen W, Zhang M, Huo B. The effects of trust and coercive power on supplier integration.

International Journal of Production Economics. 2009;**120**(1):66-78

[52] Peter Dapiran G. Are co-operation and trust being confused with power? An analysis of food retailing in Australia and the UK. International Journal of Retail and Distribution Management. 2003;**31**(5):256-267

[53] Corsten D, Kumar N. Do suppliers benefit from collaborative relationships with large retailers? An empirical investigation of efficient consumer response adoption. Journal of Marketing. 2005;**69**(3):80-94

[54] Kahkonen AK, Virolainen VM. Sources of structural power in the context of value nets. Journal of Purchasing and Supply Management. 2011;**17**(2):109-120

[55] Dahl RA. The concept of power. Behavioral Science. 1957;**2**(3):201-215

[56] Wrong DH. Some problems in defining social power. The American Journal of Sociology. 2007;**73**(6):673-681

[57] Crook TR, Combs JG. Sources and consequences of bargaining power in supply chains. Journal of Operations Management. 2007;**25**(2):546-555

[58] Ireland RD, Webb JW. A multi-theoretic perspective on trust and power in strategic supply chains. Journal of Operations Management. 2007;**25**(2):482-497

[59] Thomas R, Sargent LD, Hardy C. Managing organizational change: Negotiating meaning and power-resistance relations. Organization Science. 2011;**22**(1):22-41

[60] Anderson E, Weitz B. Determinants of continuity in conventional industrial channel dyads. Marketing Science. 1989;**8**(4):310-323

[61] French J, Raven B. The bases of social power. In: Cartwright D, editor. Studies in Social Power. Ann Arbor: University of Michigan Press; 1959. pp. 150-167

[62] Mintzberg H. Le pouvoir dans les organisations. Paris: Editions d'organisation; 2003

[63] Ansari MA, Kapoor A. Organizational context and upward influence tactics. Organizational Behavior and Human Decision Processes. 1987;**40**(1):39-49

[64] Bacharach SB, Lawler EJ. Power and tactics in bargaining. Industrial and Labor Relations Review. 1981;**34**(2):219-233

[65] Sotarauta M. Power and influence tactics in the promotion of regional development: An empirical analysis of the work of Finnish regional development officers. Geoforum. 2009;**40**(5):895-905

[66] Rojot J. La Négociation. Paris: Vuibert; 2006

[67] Li B, Zhou Y, Wang X. Equilibrium analysis of distribution channel structures under power imbalance and asymmetric information. International Journal of Production Research. 2013;**51**(9):2698-2714

[68] Ford RC, Wang Y, Vestal A. Power asymmetries in tourism distribution networks. Annals of Tourism Research. 2012;**39**(2):755-779

[69] Wang CH. The moderating role of power asymmetry on the relationships between alliance and innovative

performance in the high-tech industry. Technological Forecasting and Social Change. 2011;**78**(7):1268-1279

[70] Kim PH, Pinkley RL, Fragale AR. Power dynamics in negotiation. The Academy of Management Review. 2005; **30**(4):799-822

[71] Cook K, Emerson R, Gillmore M, Yamagishi T. The distribution of power in exchange networks: Theory and experimental results. The American Journal of Sociology. 1983;**89**(2): 275-305

[72] Emerson R. Power dependence relations. American Sociological Review. 1962;**27**(1):31-41

[73] Brass DJ, Burkhardt ME. Potential power and power use: An investigation of structure and behavior. The Academy of Management Journal. 1993;**36**(3): 441-470

[74] Lee S. The impact of office information systems on potential power and influence. Journal of Management Information Systems. 1991;**8**(2):135-151

[75] Molm LD. The conversion of power imbalance to power use. Social Psychology Quarterly. 1981;**44**(3):151-163

[76] Spangler B. Integrative or Interest-Based Bargaining. Technical Report. Beyond Intractability, Guy; 2003

[77] Ford R, Johnson C. The perception of power: Dependence and legitimacy in conflict. Social Psychology Quarterly. 1998;**61**(1):16-32

[78] Hedstrom P, Swedberg R, Udehn L. Popper's situational analysis and contemporary sociology. Philosophy of the Social Sciences. 1998;**28**(3): 339-364

[79] Chaneta DI. Role of situation analysis in strategy-making. Journal of Comprehensive Research. 2008;**6**(1): 39-46

[80] Liker J. The Toyota Way: 14 Management Principles from the World's Greatest Manufacturer. 1st ed. New York: McGraw-Hill Education; 2004

[81] Fine CH. Clockspeed: Winning Industry Control in the Age of Temporary Advantage. New York: Basic Books; 1998

[82] Hill T, Westbrook R. SWOT analysis: It's time for a product recall. Long Range Planning. 1997;**30**(1):46-52

Chapter 2

Driving Operational Resilience: Perspectives from Emerging and Developing Economies

Marvel Ogah

Abstract

Some organisations have supply chain ecosystems domiciled in Africa and have yet to fully recover from current global disruptions' impacts. A critical aspect of any supply value stream is the ability to recover from internal and external shocks or vagaries that may affect its ability to provide value incrementally to its customers. Another term for this attribute is operational resilience or supply chain resilience. The chapter discourse will include the concept of supply chain resilience, incidents of supply chain disruptions, global best practices for eliciting operational resilience, and key recommendations for key stakeholders.

Keywords: Africa, developing economies, operations, resilience, supply chain

1. Introduction

Globally, most organisations have had to contend with myriad of issues bothering on the sustenance of their operations after recent or immediate disruptions. Some organisations have been able to recover from this level of disruption, some are yet to do, and some have had difficulty in attaining this feat, as it were. Resilience, from the perspective of operations or supply chain, refers to the ability or propensity of an organisation's supply chain or operation's architecture to proactively elicit value-add in alignment with customer requirements due to the vagaries of dynamic environment. Thus, organisations harness the capability to continuously innovate and retool their preparedness and response towards an unstable and changing business world [1, 2]. This situation requires a continuous re-jigging and re-tooling of business processes and supply flows geared towards the need to suit and engage the disruptive and unstable external forces in tandem with the capability to rebound to normalcy.

2. Concept of resilience

Concept of resilience is somewhat amorphous and ambiguous because it draws its meaning from diverse backgrounds such as ecology, engineering, psychology, operations, and supply chain management. However, the concept of resilience could be captured within the domain of supply chain in tandem with its original definition. Thus, according to Rodin [3], resilience could be defined as the capacity of an entity—an individual, a community, an organisation, or a natural system to prepare for disruptions, to recover from shocks and stresses, and to adapt and grow from a disruptive experience. The definition of resilience as a concept can be adapted from the dimensions of physical, psychological, economic, engineering, disaster management, organisational, ecological, and socio-ecological [4–8]. This school of thought defines resilience as the ability of a system to return to its original state or move to a desirable state after an incident of disturbance and affected by an external event or action. In other words, it is the capacity of an entity or enterprise to survive, adapt, and grow amidst turbulence; it also, the inherent and latent ability of a system to endure despite exposure to some measure of adversity, recover, and maintain its existing structure after a shoch as a function of its flexibility, and stability. This perspective embodies the notion that organisations seek stability and equilibrium towards attaining efficiency amidst disruption; this perspective takes cognisance of eliciting responsive measures, risk mitigants, and eliciting capability as way of overcoming vulnerabilities. These definitive perspectives of resilience have subsisted over the years across extant literature and practice; however, there exist some contradictory stance from some schools of thoughts regarding how organisations can leverage their supply chain architecture to build or enhance resilience capability quotient [6, 8, 9]. Thus, a school of thought thinks that resilience capabilities encompass flexibility, redundancy, collaboration, visibility, agility, and multiple sourcing, while another of school of thought thinks that resilience entails capacity, culture, inventory, information sharing, flexibility, and responsiveness. With recent upheavals from the incidents of global disruption, this aspect of resilience has dovetailed to considering supply chain network design through configuration as a critical dimension of resilience capability. Aside these definitions and nomenclatures of resilience, due to inconsistency and incoherence amongst scholars and practitioners alike regarding the capability dimensioning of resilience with reference to operations and supply chain, within the context of developing economies, there is an exigent need, as it were, to explore the factors that are crucial and essential for current risk response measures and future risk mitigants in developing economies. In an alignment with this premise, there is an operational need to redesign the supply chain infrastructure in developing countries towards forming strong networks; this is necessary because it would stimulate normative and coercive pressures especially for intra and inter-organisational network with a view to overcoming existing gridlocks in organisations that are contending with operational and supply resilience issues [10–13]. Every supply chain architecture provides a relational link between a buyer and supplier network, hitherto this link has been buffeted by recurring incidents of disruption as occasioned by global geopolitics and similar occurrences, and thus, there is a need to factor in the resilience equation, the social capital perspective of resilience vis-a-vis the implication for developing countries [14, 15].

Accordingly, leveraging these perspectives, the concept of resilience has enlarged its connotation because the need to build resilience has taken an urgent social and economic dimension as function of global disruption as occasioned by urbanisation, climate change, and globalisation ([3], p. 4). These three facets are largely connected and flow into an ecological cum economic nexus. Inter alia, resilience has five characteristics [3], viz.: aware, diverse, integrated, self-regulating, and adaptive; these characteristics are subsumed within three phases of readiness, responsiveness, and revitalisation. Today's business milieu has been affected by incidents of turbulence and volatility, and these have also created unexpected shocks and discontinuities that have impacted operations and supply chain ecosystems to an unmeasurable level of disruptions. These aspects have led to considering resilience largely from the angle of supply chain.

3. Supply chain resilience

A phenomenon whereby the link between a buyer and supply is disrupted by external or internal occurrences may warrant a level of resilience otherwise known as supply chain resilience. Various definitions or conceptualisation of supply chain resilience abound. However, amongst the myriad of these definitions, supply chain resilience is the capability of a supply chain ecosystem to proactively design and plan its supply chain network with a view to anticipating unexpected disruptive occurrence geared towards responding adaptively to occasions of disruptions [16, 17]. Thus, supply chain resilience entails the ability of a supply chain flow to ramp up its anticipation capability with a bid to prepare, respond, and recoup from incidents of disruptions while maintaining service continuity at acceptable predefined levels. Globally, supply chain networks are becoming complex, interconnected, and vulnerable to a myriad of risks ranging from natural disasters to cyberattacks and global pandemics. These emerging aspects have necessitated the need for supply chain ecosystem to dovetail into having supply chains that have several strategies for managing and eliciting recovery from incidents of disruptions. Since, resilience is an embodiment of a complex, collective, and adaptive capability of organisations in the supply network positioned to maintain a dynamic equilibrium, geared towards recovery from disruptions, and regain functionality by overcoming negative impacts, responding to unexpected changes, and leveraging accumulated knowledge base to drive incremental improvement [18]. Albeit interaction and co-evolution between different organisations on one hand and between organisations and environment, on the other hand, determine and influence the robustness of the resilience hygiene to an extent as required by the business milieu. Resilience transpires with the supply value chain at three levels: firm, supply chain architecture, and supply network; these levels are fraught with some measures of uncertainty, and there is always a strategic need to drive efficiency and productivity from these dimensions [19]. These levels of uncertainty are occasioned by incidents of international terrorism, earthquakes, fires, accidents, tsunamis, and hurricanes; thus, there is an evolving emergent need to build the resilience quotient into the various strata of supply chain networks of concerned organisations. For some organisations in this category, it might be necessary for them to have control of its entire logistics flow as an important subset of its supply chain ecosystem because it can leverage on it as competitive advantage, both in the medium-, and long-term basis. For these organisations to orchestrate these as a competitive advantage in their resilience quotient, there is a need to aggregate and recalibrate some key elements of supply chain resilience into their organisational operational excellence quotient. These would be discussed as follows:

3.1 Risk identification and management

A crucial aspect of supply chain resilience bothers on the identification and management of risk and its mitigation. Modern global supply chain ecosystems are vulnerable to a gamut of risks encompassing natural disasters, political instability to pandemics, and inclusive of labour crises. Thus, organisations need to do a reconnoitre regarding how to develop risk mitigation plans towards preparing, responding, and recovering from disruptions by examining the interplay between different capability facets of its supply chain networks. In so doing these organisations would be able to build resilience capability by strengthening individual capabilities and orchestrating them effectively. The first step for organisations to achieve this concretely resides in their abilities to identify potential risks and vulnerabilities within and along the supply chain flow. This step would entail a thorough assessment of all elements of the supply chain, as a function of suppliers, logistics, information systems, and dependencies.

3.2 Visibility and transparency

Visibility and transparency are vital aspects of any supply chain ecosystem. An aspect of any supply value chain that easily indicates symptoms of break-down vis-a-vis its resilience is an a lack or convoluted visibility coupled with a tandem lack of transparency. Thus, the need to have optimal visibility across all tiers of any supply chain flow is of vital importance. Visibility entails knowledge about the sources of components, flow, conversion process, and interaction of the components within the supply chain network. On the other hand, transparency involves the identification of bottlenecks and areas of weakness that are susceptible to disruption along the supply chain flow.

3.3 Flexibility and adaptability

The sourcing and channelling of inputs into the supply value chain is susceptible to a vagary of conditions that are both external and internal. Thus, there is an emergent need subsumed from the aspect of operational excellence that resilient supply chains are flexible, agile, and nimble as a subset of any supply chain architecture to withstand or combat any incident of resilience. These intricacies would then involve the need for organisations to have alternative sources of supply, flexible logistics, and flexible production and capacity capabilities.

3.4 Collaboration and communication

Collaboration, from an immersive perspective, is the new order in the current global economy, and supply chain being an important part of this global narrative, cannot be excluded. Thus, there is a need to build and forge strong relationships amongst suppliers, partners, and organisations as part of the supply chain network to improve information flow and trust as enablers for agile responses to disruptions. A critical attribute of this dimensions is leveraging effective communication towards curtailing crisis situations, as the needs arise.

3.5 Redundancy

Building buffers, lags, and redundant subunits within any supply chain flow provide enablers for mitigation tactics. These measures would entail building inventory and backup logistics quotients in the supply chain ecosystem; these might be costly in the short-term but would provide a bucket for an insurance policy against incidents of disruptions. However, the decision to locate where to position such redundancy may involve some strategic planning and cost–benefit analysis.

3.6 Recovery planning

The stimulation of stress test and scenario planning are important aspects of recovery planning. These dimensions would elicit agile strategies for an efficient supply chain recovery after an incident of disruption. Recovery planning would also entail re-calibrating the operational assumptions that undergird the resilience quotient within the supply chain ecosystem.

3.7 Continuous improvement

The incidents of disruptions that may warrant need to build resilience into any supply chain ecosystem are pervasive. To this end, organisations operating globally, with reference to both developing and developed economies, should be in a state of readiness, as default, to mitigate disruption, because building supply chain resilience is not a one-off achievement, but it should be heralded on a continuous basis. Thus, this would entail regular reviews and updates to strategic intents geared towards mitigating the incidents of disruption via the resilience mandate; these measures would then involve incorporating lessons learnt from previous incidents of disruptions and keeping abreast of emerging technologies and methodologies towards maintaining resilience.

4. Incidents of supply chain disruptions in developing economies

Incidents of supply chain disruptions abound in developing economies coupled with dire consequences being exacerbated by week infrastructural architectures and unstable economic policies. However, navigating the resilience quotient is fraught with some dimensions that pose long-lasting challenges of social sustainability and novelty with the need for global restoration vis-a-vis operational stability [20, 21]. These dimensions have unique operational implications for governments and stakeholders in these developing countries. For developing countries in sub-Saharan African, there exists frequent challenges of disruption in their supply chain networks affecting both local economies and global markets. These challenges have occasioned significant supply chain disruptions in these regions:

4.1 Infrastructural deficiencies and deficits

Most developing countries in sub-Saharan African have contended with issues of infrastructural deficiencies in some cases, and other case, outright deficits. For instance, most rural areas of developing countries are inaccessible by reliable roads, which have become a burden to the supply chain network, and these situations complicate transportation of goods and drive incremental costs. Also, these roads become impassable during rainy seasons, leading to significant delays. Another aspect of this challenge is the sporadic and instability in energy supply as occasioned by frequent power outages, which negatively impacted manufacturing and other industrial operations, leading to operational inefficiency cum delays, and increased operational costs. These infrastructural deficiencies and deficits as occasioned by inadequate roads, poor rail networks, and coupled with epileptic power supply have resulted in high operational costs, affecting the efficiency and reliability of supply chains, and ultimately disrupted supply chains.

4.2 Political instability and conflict

Political instability and conflict are characteristic features of business environments in developing economies, and these portend a substantial vulnerability on the sides of most business operations in these countries. Instances of these occurrences abound in these countries. Nigeria has witnessed incessant incident of insurgency in the middle and northern parts and in the Niger Delta areas; these have impacted supply chain operations of manufacturing, service, oil production, and retail businesses in these areas. Insurgencies in the northern parts of Nigeria have also affected food supplies for human consumption and Agro in-feeds for manufacturing. These incidents require the intervention of resilience calibration because of their impact of increasing cost of foodstuff and global prices of oil in these economies. In southern Sudan, there has been a civil war which started in 2013, and it has persisted hitherto with dire consequences for humanitarian logistics and livelihood in this country. This unending conflict has impacted and disrupted food supply chains and agricultural ecosystems in this country leading to severe logistics breakdown and humanitarian crisis with a ripple effect of millions of persons displaced and in dire need of food and medication.

4.3 Economic factors

Another sphere where disruption has affected the supply chain and operations in the developing economies significantly is economy. Countries such as Nigeria, Ghana, Zimbabwe, and South Africa have witnessed significant devaluations of their currencies, which have negatively impacted the cost of livings, hike in prices of foodstuffs, consumables, and services, and leading to prices fluctuations and instabilities across local markets in these regions. This situation has constituted a bane for a significant populace and businesses in these countries. In some cases, like Nigeria, the scale of conflict and violence have resulted in perennial occurrences of kidnapping, banditry, lost of lives, injuries, communal conflicts, acts of terrorism, and ultimately resulting in a severe disruption in the food supply chain and an exacerbated food insecurity in the region. Some areas are yet to navigate their exits from the economic comatose hitherto, and it rather seems that these incidents have defied logic for leadership cadres in these countries, especially in sub-Saharan Africa. Aside from currency devaluation, existence of trade barriers in some of these countries has also warranted the need for operational resilience. Existence of high tariffs and non-tariffs has, hitherto, constituted bottlenecks impeding the free flow of good across the borders between these countries. Within the ECOWAS region, leaders and governments in these regions are hopeful that initiatives such as the African Continental Free Trade Area (AfCFTA) will have a leverage to resolve these economic bottlenecks. According to Mui and Muthuveloo [22], business environmental challenges that cause changes in business processes and practices require continuous calibration of operational flow geared towards unravelling disruptive quagmire in a bid to decouple challenges in regulatory cum economic measures that underpin operational practices of organisations. Thus, there is an evolving need to build resilience into the supply chain ecosystems of these economies as means of mitigating the impact of inflation, currency devaluation, and deficit of financial services which have disrupted the availability of goods and services in these regions.

4.4 Natural disasters and climate impacts

From 2019 to 2020 and early 2021, large swarms of locust significantly affected the agricultural value chains in Ethiopia, Kenya, and Somalia; these incidents occasioned the disruption of the food supply chains in these regions impacting severely the food security in the agriculture ecosystem. Hitherto, some of these regions, though, have deployed some level of response to these incidents, and substantial resilience architecture is yet to re-calibrated and instituted in these regions. Incidents of droughts and floods have also affected food production and supply chains in Southern and Eastern Africa.

4.5 Health and dire humanitarian crisis

The occurrence of health and humanitarian crisis in the form of epidemic has impeded the reliability and availability of labour and workforce needed to activate the supply chains, locally, and globally, in some instances. These types of incidents have affected the service-based organisations because labour-hours are essential for value delivery in these regions. Incidents of HIV / AIDS pandemic are some examples of these incidents. In countries such as Democratic Republic of Congo, populace domicile in these regions has experienced the plague of Ebola outbreaks which have affected human lives and disrupted the availability of manpower to provide important health service, the flow of quarantine measures, and outright scarcity of vaccines and medications. On a larger scale, epidemics and pandemics have severely impacted supply chain with the aftermath of workforce shortages and movement restrictions of goods and persons. For instance, in West Africa, from 2014 to 2016, there was an outbreak of Ebola which impacted supply chain flows, movement of goods and people, with a tandem effect on availability of health services and vaccines.

4.6 Regulatory and bureaucratic inefficiencies

Another cankerworm that has precipitated the need for operational and supply chain resilience in the developing economies is the existence of a myriad of regulatory and bureaucratic inefficiencies. This is a major bane for most leaders, key players, and governments in this region. Ganin et al. [23] asserted that resilience is hinged on the level of resources and rapidity flow developed as a function of an aggregated response to emerging profane issues in a supply chain stream positioned to help a system recover from swerving from its course and with minimal operational failure consequences. Thus, in most ports in Nigeria, Ghana, and Kenya, inefficient customs' procedures have constituted unending delays and bottlenecks resulting in incessant incidents of delays, and unnecessary bureaucratic constraints, where processes are usually slow and unpredictable causing prohibitive of demurrage, and operational costs.

5. Strategies to mitigate and build resilience for developing countries

Efficient and optimal supply chain networks should be in states of readiness to response substantially to business environmental changes that may impact on their operational flows of value-add [1]. To this end, it behoves organisations and governments operating the supply chain ecosystem to explore strategic options regarding how these organisations can foster and build resilience in their sub-Saharan Africa supply value chains. However, this strategic option is not only open for developing economies like Nigeria, Ghana, Zimbabwe, Cameroun, Kenya, and others, and it can be explored by countries in the global markets that use regional resources and products for their operations; this approach would not only elicit the requisite resilience from the angles of operations and supply chain, but it would also help them to enhance their regional stability and economic growth. Not withstanding, supply chain disruptions warehoused in developing economies have had tripartite impacts on businesses, economy, and society; however, characteristics of these disruptions may be influenced by socio-economic, political, and environmental factors that tailor-made for specific regions. Noteworthy, incidents of supply chain disruptions in developing countries or economies have unique features that require special solutions for mitigating the incidents and types of disruptions in this region as compared to other parts of the world. Due to the level of severity and scale of the disruptions experienced in these developing countries, practitioners, supply chain leaders, and governments can leverage some of these strategies to build or infuse resilience into their supply chain ecosystems within the sub-Saharan Africa:

5.1 Diversification of the supply chain base

The need to have to some level of diversification within supply chain networks has become a business imperative. In a rapidly evolving ecosystem, diversification as a function of agility, window for options, and responsiveness is critical for leveraging resilience. According to Rodin [3], diversification entails system redundancy; it requires a backup in a network of flows or operational architecture positioned for the avoidance of cascading system failure or dysfunctionality during crises or an incident of disruption. Sub-Saharan African countries need to have backups for their supplies via multiples sources of supplies and nearshoring their supply bases in multiples locations or collaborations amongst close neighbouring countries to mitigate the impact of unforeseen incidents of disruptions.

5.2 Leveraging technology as an enabler

A critical game changer in world that is rapidly immersed in a state of flux is having recourse to technology as an enabler for value delivery for organisations' customers. Developing operational resilience with the aid of technology is about managing risk and resilience of extended supply chain networks with some algorithms that go beyond the traditional continuity in business and disaster recovery capability but requires a rejigging of organisational capability with the prowess of technology; this ability would provide a leeway that would enable a 360-degree awareness about pending risks and mitigation tactics for eliciting resilience. African countries can leverage technologies such as Internet of things (IoT), AI, and blockchain to enhance visibility, predictive capabilities, and data analytics to build some level of operational resilience and efficiency in their supply networks.

5.3 Optimal inventory management

Inventory in the forms of raw materials, consumables, machine parts, work-in-progress (WIP), and finished are the basic blocks of any supply value chain. From the food needs for humans, medicines, to raw material for factory operations, an emergent dimension of inventory impact is the need to move from "just-in-time" to "just-in-case" for organisations in sub-Saharan African countries to have a buffer stock of critical inventory and stock stocks mitigate the severe impact of stock-out of critical inventory items that might affect their operations. This is even more critical for those organisations that have their supply chain networks in these developing countries that are largely susceptible to global economic vagaries.

5.4 Capacity building

In recent times, several developing countries in sub-Saharan Africa have had their supply chain operations witnessed some level of disruptions due to incident of manpower losses. These losses have been occasioned by talented personnels migrating to the developed countries due to emerging job opportunities in these countries; a resultant outcome of this situation is depletion in the pool of talent or capable manpower that would be needed for the overall well-being of the supply chain ecosystem in these developing countries. Countries such as Nigeria, Ghana, Cameroun, and Kenya have had a severe hit from this aspect of talent losses, and this has caused some level of disruption in their supply chain ecosystem.

5.5 Scenario planning

A proactive approach for government, practitioners, and leaders in the supply chain ecosystems is to leverage some elements of stress test and scenario planning as strategic options for mitigating the impact of disruptions on their operations. These options would enable them to regularly instigate and orchestrate scenario and simulation operational exercise that would help to test their level of agility and responsiveness to incidents of disruption; this approach would dovetail to explore viable options from a pragmatic dimension the art of recalibrating their business assumptions towards building a robust resilient quotient in their operations.

6. Global best practice for eliciting operational resilience

Several global best practices abound for eliciting operational resilience. Some of these practices can be curated from the strategies earlier discussed in the preceding section; however, with a focus on developing countries or developing economies in sub-Saharan African as a basis for this chapter, governments and key stakeholders can have recourse to the following: substantial investment in infrastructure, regional cooperation, stabilising their local currencies via stimulating their productive capabilities as a way of using more of their local products for consumptions, enhancing their service and agricultural sectors as alternative sources of income and inputs, and stabilising their political and economic environment to foster in-flow of investments, develop local alternatives as inputs, close the gaps in their infrastructural deficits by leveraging more regional collaborations and alliances such ECOWAS and AfCFTA.

7. Key recommendations for policymakers and key stakeholders

Some recommendations for policymakers and key stakeholders about building resilience in the operational and supply chain networks would include:

1. Develop local capability from the perspectives of manpower and alternate source of inputs. A critical game changer for key stakeholders, governments, and policymakers is to explore how to develop a pipeline of sustainable talents and capability within the supply value chain; though this is a necessity for both developed and developing economies, it is more crucial for most developing economies at this period of unpredictable disruptive occurrences.

2. Enhance local productive base so as to reduce over-dependence on imported items. A major issue of struggle for most developing countries is how to source materials to drive value creation; a huge chuck of critical raw materials and consumables are imported via ports that are saddled with epileptic infrastructures and graft, these in-turn make the importation of items to be expensive, and thus, there is an urgent need to enhance local productive base.

3. Backward integration as a measure for building capacity of local suppliers. This option would provide a leeway for building resilience via sourcing local suppliers.

4. Recalibration of policy and business framework. Unstable business framework and policy framework are a bane in most developing countries, and these situations stifle resilience to a large extent; thus, there is a need for governments and key stakeholders to question the assumptions undergirding these policy frameworks with a view to building some measure of flexibility and elasticity; these resonate with resilience amplification.

5. Bolstering regulatory framework. The need to build a robust resilience quotient aligns with the bolstering of regulatory framework that allows for control with some measure of flexibility. Tight controls amidst graft and lack of transparency; this is at variance with resilience.

6. Leveraging innovation and technological prowess. Innovation and technological prowess are enablers to retool resilience; most developing countries are in dire need of both to elicit resilience.

7. Retooling infrastructural framework for adequacy and fitness for purpose. The deficits in infrastructural framework are re-occurring decimals in most developing countries such Nigeria, Ghana, and Kenya; this situation has trapped the capability to unleash resilience within most value chains in these developing economies.

8. Conclusion

Resilience has been a synonym for operational relevance and survival in business environments that are fraught with unending and unplanned upheavals. It is the required business elixir for survival, relevance, and sustainability in an evolving global business environment that is characterised by massive indices of disruptive tendencies of volatility, uncertainty, complexity, and ambiguity. To this end, organisations that want to thrive in sub-Saharan supply chain ecosystems should endeavour to consider the resilience capability as requisite option to remain relevance within the global ecosystem. This approach would necessitate an understanding of threats or disruptive tendency as a deviation from the resilience equilibrium. In doing so, this chapter's discourse explored the concept of resilience from the framework of supply value chain, supply chain resilience, and incidents of supply disruptions in developing economies. Aside these, the discourse indicated some global best practices, key recommendations that organisations and governments domicile in these countries need to leverage as viable options for eliciting resilience. Thus, it is essentially crucial that governments, leaders, operators, and stakeholders build sustainable and resilient supply chain networks that can withstand global and local shocks. This approach would require a consolidated dimension that would balance cost with benefits for all players and stakeholders in the supply value chain and create business enablers that would also elicit resilience as competitive advantage and a measure of reliability and trustworthiness.

Author details

Marvel Ogah
Lagos Business School, Pan-Atlantic University, Lagos, Nigeria

*Address all correspondence to: mogah@lbs.edu.ng

References

[1] Essuman D, Boso N, Annan J. Operational resilience, disruption, and efficiency: Conceptual and empirical analyses. International Journal of Production Economics. 2020;**229**:107762

[2] Ufua DE, Itai M, Kumar A, AI-Faryan MAS. Achieving operational resilience through kaizen practice: a case in a commercial livestock farm in Nigeria. The TQM Journal. 2023;**36**(4):1092-1112. DOI: 10.1108/TQM-01-2023-0013

[3] Rodin J. The Resilience Dividend: Being Strong in A World Where Things Go Wrong. New York: Public Affairs; 2014

[4] Christopher M, Peck H. Building the resilient supply chain. The International Journal of Logistics Management. 2004;**15**(2):1-13

[5] Fiksel J. Sustainability and resilience: toward a systems approach. Sustainability: Science Practice and Policy. 2006;**2**(2):14-21

[6] Sajko M, Boone C, Buyl T. CEO greed, corporate social responsibility, and organisational resilience to systemic shocks. Journal of Management. 2021;**47**(4):957-992

[7] Svensson G. A conceptual framework of vulnerability in firms' inbound and outbound logistics flows. International Journal of Physical Distribution and Logistics Management. 2002;**32**(2):110-134

[8] Pettit TJ, Croxton KL, Fiksel J. The evolution of resilience in supply chain management: A retrospective on ensuring supply chain resilience. Journal of Business Logistics. 2019;**40**(1):56-65. DOI: 10.1111/jbl.12202

[9] Pettit TJ, Fiksel J, Croxton KL. Ensuring supply chain resilience: Development of a conceptual framework. Journal of Business Logistics. 2010;**31**(1):1-21. DOI: 10.1002/j.2158-1592.2010.tb00125.x

[10] Butt AS. Strategies to mitigate the impact of COVID-19 on supply chain disruptions: a multiple case analysis of buyers and distributors. The International Journal of Logistics Management. 2021;**35**(7):0957-4093. DOI: 10.1108/IJLM-11-2020-0455

[11] Paul SK, Chowdhury P. Strategies for managing the impacts of disruptions during COVID-19: an example of toilet paper. Global Journal of Flexible Systems Management. 2020;**21**(3):283-293

[12] SonBG,ChaeS,Kocabasoglu-HillmerC. Catastrophic supply chain disruptions and supply network changes: a study of the 2011 Japanese earthquake. International Journal of Operations and Production Management. 2021;**41**(6):781-804

[13] Jajja MSS, Kannan VR, Brah SA, Hassan SZ. Linkages between firm innovation strategy, suppliers, product innovation, and business performance: Insights from resource dependence theory. International Journal of Operations & Production Management. 2017;**37**(8):1054-1075. DOI: 10.1108/IJOPM-09-2014-0424

[14] Lang LD, Behl A, Dong NT, Thu NH, Dewani PP. Social capital in agribusiness: an exploratory investigation from a supply chain perspective during the COVID-19 crisis. The International Journal of Logistics Management. 2021;**33**(4):1437-1473. DOI: 10.1108/IJLM-01-2021-0039

[15] Shivarajan S, Srinivasan A. The poor as suppliers of intellectual property: a social network approach to sustainable poverty alleviation. Business Ethics Quarterly. 2013;**23**(3):381-406

[16] Aghai-Khozani H, Bull S, Dilda V, Mori L, Reiter S. A more resilient supply chain from optimized operations planning. Operations Practice. 2022. Available from: https://www.mckinsey.com/capabilities/operations/our-insights/a-more-resilient-supply-chain-from-optimized-operations-planning

[17] Lopes JM, Gomes S, Mané L. Developing knowledge of supply chain resilience in less-developed countries in the pandemic age. Logistics. 2022;**6**(3). DOI: 10.3390/logistics6010003

[18] Yao Y, Fabbe-Costes N. Can you measure resilience if You are unable to define it? The analysis of Supply Network Resilience (SNRES). Supply Chain Forum: An International Journal. 2018;**19**(4):255-265. DOI: 10.1080/16258312.2018.1540248

[19] Kim Y, Chen Y-S, Linderlan K. Supply network disruption and resilience: A network structural perspective. Journal of Operations Management. 2015;**1**:33-34. DOI: 10.1016/j.jom.2014.10.006

[20] Ivanov D. Predicting the impacts of epidemic outbreaks on global supply chains: a simulation-based analysis on the coronavirus outbreak (COVID-19/SARS-CoV-2) case. Transportation Research. Part E, Logistics and Transportation Review. 2020:136

[21] Silvestre BS. Sustainable supply chain management in emerging economies: environmental turbulence, institutional voids and sustainability trajectories. International Journal of Production Economics. 2015;**167**:156-169

[22] Mui KS, Muthuveloo R. The moderating effect of kaizen culture on the relationship between innovation and operational performance. In: 8th International Conference of Entrepreneurship and Business Management Untar. 2020. pp. 12-18. Available from: https://www.researchgate.net/publication/342774599

[23] Ganin AA, Massaro E, Gutfraind A, Steen N, Keisler JM, Kott A, et al. Operational resilience: concepts, design and analysis. Scientific Reports. 2016;**6**(1):1-12

Chapter 3

Inventory Management

Ahmed Esmail Mohamed

Abstract

Inventory management is pivotal for business performance and profitability. Efficiently handling inventory levels enhances operational efficiency and financial strategies. Poor management can result in financial losses, stock imbalances, delayed order fulfillment, and dissatisfied customers. This chapter discusses inventory management frameworks, focusing on objectives, techniques, and best practices. The primary goal is to balance overstocking and understocking, ensuring adequate working capital while optimizing costs. The chapter delves into various inventory management techniques, aiming to effectively control and manage inventory. It particularly sheds light on two pivotal techniques: Vendor-Managed Inventory (VMI) and Just-in-Time (JIT). VMI reduces inventory levels and improves fill rates, enhancing supply chain performance. On the other hand, JIT minimizes excess inventory and improves material flow. Both techniques present challenges and benefits. Accurate forecasting is highlighted as a best practice, aligning production with demand, reducing carrying costs, enhancing cash flow, and streamlining the supply chain. The literature is reviewed, emphasizing inventory management's role in the broader context of supply chain management.

Keywords: supply chain management, inventory management, cost efficiency, demand forecasting, Vendor-Managed Inventory, safety stock, Economic Order Quantity

1. Introduction

Modern business management no longer competes as independent entities but rather as supply chains. Mentzer [1] defines a supply chain as "a set of three or more entities (organizations or individuals) directly involved in the upstream and downstream flows of products, services, finances, and/or information from a source to a customer". The supply chain is not only a chain of businesses with one-to-one, business-to-business relationships but also a network of multiple businesses and relationships [2].

Supply chain management (SCM) encompasses the flow of both materials and information within the supply chain, with the latter directly influencing internal activities such as production scheduling, inventory control, and delivery plans [3]. Effective supply chain management decisions can result in cost reductions for organizations, including transportation, sourcing, stockouts, storage, and disposal costs. To achieve these benefits, supply chain members must recognize their role within a complex network and understand the interdependencies between chain participants.

 IntechOpen

The overall objectives of SCM are to improve the company's profitability and performance, to reduce the costs mainly by reducing the inventory level, and to maximize the overall value generated to the customer [4].

In the context of supply chain management, the integration of inventory management is essential. Efficient inventory management ensures that the right products are available at the right time, preventing stockouts and overstocking [3]. This synchronization enables organizations to respond promptly to changes in demand while minimizing excess inventory costs. Inventory management presents numerous challenges for many supply chains as they are becoming more complex and composed of multiple stages.

The goal of inventory management is to increase customer service, enhance product variety and availability, and minimize costs [5]. Achieving efficient supply chain operations relies on well-defined inventory management policies that facilitate reduced excess inventory levels throughout the entire chain [6].

2. Literature review

This section will offer an overview of inventory management aspects that have been identified in prior literature. Research on inventory management is crucial and has consistently held a central role in various academic literatures. Scholars from diverse fields have published articles that contribute to the advancement of inventory management theory.

Inventory is commonly understood as being synonymous with stock. Stock refers to tangible goods that are subject to processes such as mining, conversion, creation, transportation, and sale [7]. Inventories include raw materials, work-in-process goods, and fully finished products, comprising a business's assets that are either prepared for immediate sale or intended for future availability [7]. Selecting an appropriate inventory model is regarded as a significant challenge for industries. The earliest scientific research on inventory management dates back to the second decade of the previous century, yet the continued interest in this scientific field remains substantial [8]. Inventory management is defined as an ongoing procedure involving the planning, organization, and control of inventory [9]. Its objective is to reduce investment in inventory while ensuring equilibrium between supply and demand. The objective of this process is to decrease procurement and carrying costs, all the while upholding an optimal product inventory level to satisfy customer demand.

Inventory management tracks stock trends, ensuring timely ordering and customer order fulfillment while preventing shortages. Inventory transforms into revenue upon sale, but while unsold, it ties up cash, and excessive stock diminishes cash flow [5]. Effective inventory management entails weighing inventory costs against its benefits. Inventory management's primary advantages include enabling order fulfillment and increasing profits [10]. Inventory offers numerous advantages. It enhances customer service through reduced lead times, fosters organizational flexibility, cushions variances between input and output rates—accomplished by bulk ordering or manufacturing and selling in smaller quantities—and evens out manufacturing burdens for organizations dealing with demand fluctuations [7]. Additionally, it serves as a repository for potential capacity that might otherwise go to waste. Inventory also has disadvantages. It ties up working capital and space. It can suffer from obsolescence, deterioration, and shrinkage. It can lead to administrative complexity and can mask inefficiencies.

Inventory management is widely regarded as one of the most critical challenges in organizational management. Typically, there is no one-size-fits-all solution; the circumstances vary for each company or firm, encompassing distinct features and limitations [8]. In the present day, numerous inventory management techniques are employed by companies with the intention of gaining control over their inventories. However, not all these techniques are universally applicable to every industry. As a result, despite the multitude of approaches available for inventory control, selecting an appropriate one that aligns with a specific company remains a formidable challenge. Within this chapter, we will delve into several of these techniques and elucidate the significance of each.

3. Inventory management

Inventory management is a specialized branch of business management that involves the strategic planning and control of inventories, aiming to uphold the ideal stock quantities for certain products or items [11].

The challenges of inventory management revolve around the balance between excessive stock and insufficient inventory. Effective inventory management is crucial due to its direct impact on operational efficiency, customer satisfaction, and the financial objectives of an organization. As the number of stages in the supply chain increases, the complexity of managing inventory also rises. Handling inventory within a multi-echelon supply chain is notably more challenging than overseeing it within a single-echelon structure [12].

3.1 Objectives of inventory management

In businesses, especially those involved in manufacturing and production, a substantial number of financial resources are typically directed toward inventory. This allocation of funds can sometimes exceed what is dedicated to other assets within the organization. Consequently, inventory management emerges as a central player within the realm of supply chain management activities. As previously noted, the fundamental objective of inventory management encompasses ensuring an uninterrupted supply of materials to sustain production during customer demand while avoiding both overstocking and understocking. This approach aims to maintain the necessary working capital for operational and sales activities; optimize various costs such as carrying, ordering, and storage costs; eliminate duplication in stock orders; provide a continuous flow of required materials; and contribute to price stability. In essence, effective inventory management serves the critical purposes of supporting consistent production, cost efficiency, and overall financial stability. In this section, the focus will be on the main, common points that are mentioned in the previous literature and are related to the objectives of inventory management.

3.1.1 Cost efficiency

Inventory management is associated with costs that exert a direct influence on the value of inventory [13].

3.1.1.1 Carrying costs

Among the challenges of inventory management, carrying costs stand out. These costs arise from the storage of products in warehouses, distribution centers, or stores. They encompass expenses such as storage fees, labor, transportation, handling, taxes, insurance, and depreciation. In essence, carrying costs constitute the total expenditure invested in stocking and storing products and items prior to their sale.

3.1.1.2 Ordering costs

Ordering costs pertain to the expenses linked with placing orders for new inventory. This includes transportation, shipping charges, inspections, and other expenditures related to the order process.

3.1.1.3 Stockout costs

Stockout costs encompass both direct and indirect expenses that a business incurs when it faces a shortage of stock. Recognizing the substantial impact of these costs on a company's margins, revenue, and overall profitability is crucial.

3.1.2 Optimal inventory levels

Balancing on-hand stock is a central challenge faced by most businesses. Achieving optimal stock levels necessitates alignment with actual customer demand, ensuring that quantities are consistently available to meet that demand. Simultaneously, maintaining excessive stock levels can lead to a host of adverse consequences that negatively impact cash flow and profit margins.

3.1.3 Demand forecasting

Another main challenge for inventory management is demand forecasting, which involves anticipating inventory needs. Supply chain forecasting goes beyond the operational task of estimating demand requirements at each level of the supply chain network. It involves complex issues such as supply chain coordination and sharing end-customer demand information among members. Consequently, numerous forecasting techniques have been investigated to assess their impact on inventory management in the supply chain. Researchers have extensively examined methods such as moving averages, exponential smoothing, and minimum mean squared error. Demand forecasting plays an important role in balancing inventory levels and consequently affects inventory management.

3.1.4 Lead time management

Lead time represents the time difference between when a customer places an order and when the product or service is delivered. It serves as a crucial factor for calculating demand during the lead time, also known as safety stock [3]. In short, lead time is the duration it takes for an order to be fulfilled. Lead time can significantly impact the decision-making process when purchasing products, as longer lead times can cause companies to become apprehensive and order excessive amounts, ultimately leading to an accumulation of excessive stock.

3.2 Inventory management techniques

Inventory management is an immensely significant function for any business; if not done well, it can cause serious problems. Poorly managed inventories can lead to production delays, unhappy customers, and less available money for the business. Inefficient inventory management can result in production delays, customer dissatisfaction, or even a reduction in working capital [14].

Many researchers have looked into different ways to manage inventory effectively. In this chapter, we'll concentrate on the techniques most frequently discussed in the literature.

3.2.1 ABC inventory classification technique

The ABC inventory classification technique is widely utilized by large firms to efficiently manage a significant volume of inventory items. This method involves the categorization of inventories into three primary groups: A, B, and C. Items designated as "A" hold the highest level of importance, whereas "C" items are deemed to be of lesser significance. The classification of items follows a hierarchy where "Class A" items represent the most crucial, followed by Class B items, and finally, Class C items, which are the least critical. This classification system plays a pivotal role in prioritizing inventory items based on their stock levels and reordering requirements. Moreover, it simplifies the arrangement of inventory items based on their respective values, demand, costs, and risk data [15].

3.2.2 Just-in-Time (JIT) inventory management technique

The concept of JIT originated from the Toyota Motor Company in Japan. JIT represents an innovative approach to inventory management, fostering a supply–demand system that promotes streamlined production and aims to precisely align demand with supply [8]. JIT systems are designed to rapidly respond to demand fluctuations, eliminating the need for excessive inventory. These systems replenish inventory and trigger reorders for future resources once a predetermined minimum threshold is reached, signaled by an indicator that denotes the necessity for additional stock to fulfill current demand. The core principle revolves around delivering goods precisely as promised when an order is placed by the customer. In manufacturing, the JIT philosophy extends to a Japanese management approach that revolves around ensuring the right items of optimal quality and quantity are available in the right location and at the right time [16]. JIT inventory management is an approach whose goal is to eliminate inventories rather than simply improve inventory quality. Raw materials and work in progress are reduced to equal the quantity required for a day's production. This is achieved by reducing planning time and lead time, allowing for smaller packaging. Suppliers may be required to deliver multiple items per day or near the utility facility to support this process [10].

3.2.3 Economic Order Quantity (EOQ)

The order quantity that minimizes overall costs is referred to as the Economic Order Quantity (EOQ). This quantity strikes the most efficient balance between ordering and holding costs, making it a crucial parameter in inventory management. The concept of Economic Order Quantity (EOQ) revolves around determining the optimal order quantity that strikes a balance between inventory holding costs and reorder costs [17]. This model has proven to be invaluable in resource optimization, leading to significant cost reductions. The EOQ model involves a careful consideration of the trade-off between ordering costs and storage costs when deciding on the replenishment quantity for item inventories.

When opting for a larger order quantity, the frequency of ordering is reduced, subsequently lowering ordering costs. However, this approach necessitates maintaining a higher average inventory, consequently elevating storage (holding) costs. Conversely, choosing a smaller order quantity reduces average inventory levels but results in more frequent orders, leading to higher monthly ordering costs.

3.2.4 Safety stock

Safety stock refers to the average amount of inventory held on hand to account for short-term uncertainties in both demand and supply variability [18]. In simpler terms, safety stock represents the inventory maintained to prevent instances of stockouts and backorders. Its purpose is to safeguard against various deviations, including discrepancies in delivery dates (when replenishment lead times vary), variations in requirements (arising from inaccurate forecasts), differences in delivery quantities (stemming from insufficient vendor supply or subpar material quality), and discrepancies in inventory levels (when disparities arise between planned and actual inventory) [19].

3.2.5 Vendor-Managed Inventory (VMI)

Vendor-Managed Inventory (VMI), also referred to as continuous replenishment or supplier-managed inventory, stands out as a pivotal partnering initiative that promotes collaboration and information sharing among trading partners, exploring the various benefits it offers [20]. The VMI technique gained prominence in the late 1980s through adoption by industry leaders such as Walmart, Procter & Gamble, Nestle, and Electrolux Italia.

Vendor-Managed Inventory (VMI) represents a strategic supply chain approach where the vendor assumes responsibility for determining optimal inventory levels for each product and establishes suitable inventory policies to ensure seamless inventory management. This approach proves advantageous to retailers by furnishing them with a competitive edge. It translates into heightened product availability and service levels while concurrently reducing the need for extensive inventory monitoring and minimizing costs associated with the ordering process.

3.3 Best practices for effective inventory management

Most retailers aim to achieve goals such as reducing inventory levels and achieving high inventory turnover through their practical operations. The outcomes of decisions and steps taken during inventory management can lead to issues like low liquidity due to a high inventory backlog or poor customer experiences resulting from inventory shortages [21]. In this section, we will delve into the key points that contribute to the best practices for effective inventory management.

3.3.1 Accurate forecasting

Accurate demand forecasting is a critical success factor in effective inventory management. It encompasses predicting, projecting, or estimating the expected demand for products over a specified future timeframe [22]. Demand forecasting plays a pivotal role in inventory management by enhancing an organization's competitiveness and facilitating informed decision-making. This, in turn, forms the basis for replenishment, distribution plans, and efficient supply chain management. Demand forecasting poses significant challenges across various industries. Two main problems make it hard. First, real customer demands can be irregular. Second, it's often unpredictable when that demand will happen. Consequently, demand forecasting becomes imperative for informed decisions regarding inventory management and production planning. By aligning inventory levels with projected demand, accurate forecasting can reduce inventory costs and ensure enhanced customer satisfaction through consistent and timely fulfillment of customer demands.

3.3.2 Supplier collaboration

Manufacturing and production companies commonly encounter a range of challenges when it comes to supplier collaboration. These challenges can be categorized into several key areas. One notable challenge is data sharing, as many companies struggle to effectively share real-time information that is crucial for informed decision-making processes. Another significant hurdle is the lack of trust between companies and their suppliers, particularly concerning sensitive information and intellectual property. Another critical aspect is Supplier Performance Monitoring, which involves the ongoing evaluation of suppliers to ensure both quality and efficiency. Nonetheless, this task can be intricate due to the varying performance metrics involved, and its implementation can be particularly challenging without the appropriate tools and resources.

To tackle these challenges, strategic planning, the adoption of advanced technologies, and the establishment of robust and open lines of communication are essential. Overcoming these obstacles often demands a substantial investment of time, resources, and commitment from all parties involved.

3.3.3 Cross-functional collaboration

Cross-functional collaboration takes place when individuals from various teams or departments collaborate on a designated project or objective. This approach harnesses the diverse skills and specialized knowledge of each team member to generate innovative ideas, tackle challenging issues, and attain superior outcomes within an organization. An illustrative example of this is evident in supply chain inventory management within many organizations. This indicates that each functional area, including sales, marketing, finance, and operations, collaborates within its specific domain of the inventory management process to align strategies and goals effectively. Cross-functional collaboration in inventory management is not solely the responsibility of one department; it involves interconnected processes that impact different parts of the business. As a result, involving different perspectives and expertise tends to improve overall inventory accuracy and align with the organization's strategic goals and objectives. This holistic approach aims to break down silos, enhance communication, and create a more cohesive inventory management strategy.

3.3.4 Regular audits

Inventory audits are a crucial process for maintaining precise stock records, identifying reasons for shrinkage, and ensuring the availability of accurate stock on hand. Such audits involve cross-referencing a company's financial and inventory records to ensure alignment with the physical inventory count, thereby ensuring inventory accuracy and addressing inconsistencies. These audits serve several critical purposes, including identifying discrepancies between actual stock quantities and accounting records, revealing overstocking or understocking scenarios to aid in asset management and inventory forecasting, and evaluating the efficiency of logistics and warehouse workflows. Ultimately, audits play an indispensable role in maintaining precise and up-to-date inventory information, contributing significantly to the overall integrity of the inventory management process.

4. Discussion

Inventory management may not be the most glamorous facet of business operations, but its importance cannot be overstated. It is a critical element in a company's profitability and growth potential. Without an effective inventory management system in place, businesses can unwittingly waste financial resources on unsellable products. Furthermore, they run the risk of experiencing stock shortages, leading to delays in order fulfillment and dissatisfied customers. These challenges can swiftly erode profit margins, presenting significant obstacles for the business.

Moreover, the consequences of inadequate inventory management go beyond financial losses and may involve the potential loss of valuable inventory items, directly impacting a company's bottom line. However, by implementing an efficient inventory management process, businesses can avoid these problems and unlock various benefits.

Expanding to the broader context of supply chain management, inventory management emerges as a pivotal component. It involves the meticulous tracking of stock levels and the movement of goods, whether for supplying raw materials to manufacturers or fulfilling orders for finished products. Inventory management is, in essence, the cornerstone of a business's longevity, offering opportunities to reduce costs, improve cash flow, and enhance profitability.

When inventory is systematically organized, it serves as the linchpin that ensures the smooth operation of the entire supply chain. Without it, a range of potential errors looms, including mis-shipments, shortages, out-of-stocks, spoilage (particularly for perishable stock items), overstocks, mis-picks, and more.

Inventory management plays a vital role in a business's overall success, profoundly influencing financial health and operational efficiency. It is a pivot upon which profitability and growth depend, demanding strategic attention and a commitment to efficiency.

In this section, our primary focus will be on two inventory management techniques that have been extensively discussed in previous literature and have demonstrated a tangible impact on effective inventory management. Additionally, our discussion will encompass the most highly regarded best practice activity that has been identified in prior research pertaining to inventory management.

4.1 Vendor-Managed Inventory

Supply chain management comprises a collection of interconnected activities aimed at achieving seamless coordination and communication within these operations, ultimately striving to enhance overall efficiency throughout the process [4, 23]. Effective supply chain strategy hinges on the sharing of product and production information among its members. It is widely acknowledged that when information sharing takes place at the retailer level, it brings about substantial benefits for the supply chain. This includes the reduction of the bullwhip effect and overall supply chain costs [3, 24].

Nevertheless, despite these compelling advantages, retailers often exhibit a reluctance to engage in information-sharing initiatives. This hesitation largely arises from the realization that the primary beneficiaries of information sharing are typically manufacturers rather than retailers themselves [3].

Numerous analytical and simulation studies have conducted extensive examinations of the factors influencing the advantages stemming from information sharing. The majority of these research investigations indicates that information sharing leads to noteworthy enhancements in supply chain performance. Furthermore, the improvements observed in the performance of supply chain enterprises through information sharing have been substantiated by a multitude of empirical studies.

As previously mentioned, Vendor-Managed Inventory (VMI), also known as continuous replenishment or supplier-managed inventory, is a highly debated partnering strategy that plays a prominent role in promoting collaboration and the exchange of information among trading partners [20]. VMI is a collaborative commerce initiative in which suppliers are granted authority to oversee the inventory of stock-keeping units on behalf of the buyer. This initiative integrates operations between suppliers and buyers through the exchange of information and the reengineering of business processes. To use VMI, buyers and suppliers need to work closely together. This means they share information and make sure their processes fit together smoothly. Typically, buyers share details about what they need and what they have in stock (information sharing). Then, suppliers take on the responsibility of managing inventory and making purchases for the buyers (process integration) [25].

Evidence has demonstrated that VMI can enhance supply chain performance by reducing inventory levels and increasing fill rates. As a result, the adoption of VMI in various industries has steadily increased over time.

By utilizing information technologies such as Electronic Data Interchange (EDI) or Internet-based XML protocols, buyers can share real-time sales and inventory data with their suppliers. Suppliers, in turn, can use this information to plan production schedules, coordinate deliveries, and efficiently manage order volumes and inventory levels at the buyer's stock-keeping facilities.

4.1.1 Challenges associated with implementing VMI

Vendor-Managed Inventory (VMI) offers mutual benefits to both suppliers and buyers. However, it presents significant challenges, including issues with data integrity, transparency, traceability, and the risk of a single point of failure due to its centralized architecture [26].

In the early stages of VMI implementation, the vendor team encounters numerous challenges, primarily centered around data collection and information gathering. Additionally, ensuring transparency and data sharing can pose significant difficulties, especially when buyers have restrictions in place [26].

Clear and well-identified communication is crucial, especially when introducing new products or discontinuing old ones. Incorporating new items into the scope can be challenging due to the absence of historical data for vendors to forecast demand. Therefore, maintaining an open line of communication between vendors and buyers is essential to ensure accurate orders, prevent stockouts, and avoid excess inventory.

Sharing data with a third party is not easily approved. Therefore, selecting a VMI partner is a challenging task, and it's advisable to consider a confidentiality agreement before implementing any collaboration.

Selecting the right VMI partner is crucial, as dissatisfaction with their services can significantly impact the supply chain. It's important to consider the possibility of ending the partnership from the beginning and have mitigation plans in place.

4.1.2 Potential benefits from VMI

Effective supply chain management is typically achieved by striking the right balance between inventory costs and customer service levels. The most desirable projects are those that yield improvements in both of these dimensions, and VMI is a prime example of such an approach [27].

Numerous studies have examined the factors that impact the advantages of VMI and other collaborative supply chain initiatives, as well as how these benefits are divided between the downstream member (buyer) and the upstream member (supplier).

Vendor-Managed Inventory (VMI) provides a competitive advantage for retailers by increasing product availability and service levels while reducing inventory monitoring and ordering costs [28]. The main advantage of VMI is its ability to generate cost savings across the entire supply chain. It streamlines inventory planning for retailers by shifting the responsibility to the vendor. This leads to a reduction in unnecessary orders and the requirement for extra storage space. With less inventory idling, carrying costs are significantly reduced.

The constant flow of data between retailer and vendor allows for more consistent and timely stock updates and orders. Unlike other supply chain management systems that rely on rough predictions, VMI uses current sales as a guide for more strategic inventory ordering. This could help in protection against the bullwhip effect. The bullwhip effect occurs when there's a misjudgment of demand for a specific product, causing disruptions in the supply chain [24]. It typically starts with retailers overestimating their demand forecasts and ordering more inventory than necessary, thereby putting pressure on suppliers to produce more of those products than actual demand warrants. Hence, VMI through information sharing can show a potential positive impact on demand and assist in the mitigation of the bullwhip effect.

Conversely, suppliers reaped numerous benefits from the adoption of Vendor-Managed Inventory (VMI). These advantages encompassed a decrease in stockouts, which not only safeguarded customer satisfaction but also resulted in increased sales. Furthermore, suppliers gained valuable insights into customer demand patterns, leading to more effective inventory planning. The capability to enhance inventory management and streamline delivery planning emerged as a noteworthy advantage for upstream members who embraced VMI [29].

4.2 Just-in-Time inventory

Over the past few decades, management has placed a stronger focus on machinery and workforce utilization while paying less attention to material flow. As a result, inventories at various stages of manufacturing were allowed to accumulate and grow without recognizing the negative impacts of this trend. Subsequently, the challenges shifted toward improving the material flow process, aiming for greater efficiency, reduced setup times, enhanced quality, and proactive maintenance. Additionally, closer coordination among different production resources, departments, and external stakeholders has become imperative.

In recent years, the term 'Just-in-Time' (JIT) and many of its synonyms, such as Zero Inventory, Stockless Production, World Class Manufacturing, and so on, have become widely recognized in repetitive manufacturing organizations worldwide [30]. It has garnered significant attention in trade and professional journals, with numerous books dedicated to exploring its various elements. Henry Ford once stated, "We have found that in buying materials, it is not worthwhile to purchase for anything other than immediate needs... If transportation were perfect and an even flow of materials could be assured, it would not be necessary to carry any stock whatsoever [31]". From this perspective, the concept of JIT began to shine.

Just-in-Time (JIT) inventory management is a strategic approach employed by businesses to receive the exact amount of inventory precisely when needed, with the primary objective of minimizing or eliminating excess inventory at any given moment. It also encompasses the goal of producing items precisely when required, in the smallest feasible quantities, and with minimal waste of human and natural resources [30]. In simpler terms, Material Requirement Planning (MRP) systems aim to achieve JIT, making it a fundamental goal for every production manager.

This approach effectively addresses the prevalent issue in commerce where excessive spending occurs on materials that are not immediately required. Its success hinges on maintaining continuous communication between the commerce company and its manufacturers, necessitating the use of specialized, streamlined software for seamless implementation.

Just-in-Time (JIT) requires a company to rely on a select few reliable suppliers. It is believed to enhance productivity and establish a leaner manufacturing system, leading to minimized inventories. This, in turn, reduces risk and helps lower the cost of manufacturing [32]. The implementation of JIT can, on average, result in reduced inventory costs, shorter lead times, and improved productivity for buying organizations.

4.2.1 Challenges associated with implementing Just-in-Time

The implementation of JIT in the purchasing of materials poses numerous challenges that can significantly impact the success of this concept. JIT techniques should be extended to suppliers, and parts and raw materials supplied by external vendors are often delivered in small batches precisely when needed. It's essential to consider that geographical and ownership factors often necessitate special provisions.

In sourcing decisions, price alone is not the sole determinant. Factors such as the long-term stability of the supplier, their ability to collaborate on problem resolution, willingness and capability to make frequent deliveries, and the quality of the parts supplied become crucial considerations [30]. Since evaluating potential suppliers requires more effort, fewer suppliers are typically used, leading to a greater prevalence of single sourcing, with existing suppliers having a competitive edge in securing additional work. This extended effort in switching suppliers also leads to the development of longer-term relationships.

Ideally, proximity to suppliers is preferred, as it facilitates easier coordination of transportation and shorter lead times. Transportation and shipping play pivotal roles in the success of JIT concepts. However, it's not impossible to apply JIT principles with remote suppliers.

4.2.2 Benefits of Just-in-Time

As previously mentioned, effective inventory management demands significant efforts to fine-tune the production system, encompassing various aspects such as methods, layout, material handling flow, workforce relationships, job content, and more. In exchange for the dedication required to implement these changes, organizations can expect to reap several benefits. The existing literature extensively outlines the advantages of JIT techniques, including but not limited to decreased inventory levels, reduced floor space requirements, shorter lead times, lower rates of scrap and rework, enhanced product quality, improved utilization of both labor and equipment, increased job satisfaction among workers, greater operational flexibility, streamlined automation, adherence to schedules, and the establishment of an alternative incentive system. All of these benefits collectively result in cost reductions, higher-quality end products, and enhanced competitiveness [16, 30, 31].

Conversely, JIT brings several advantages to the supplier. These benefits encompass a level of demand assurance, a more consistent demand pattern, and, potentially, an advantageous position when it comes to future business prospects. Nonetheless, it's important to recognize that this increased reliance on a single customer for orders can expose the supplier to heightened pressure for price concessions and other compromises [30].

4.3 Best practices under accurate forecasting

Effective inventory management is paramount for success in any business, whether in manufacturing or service industries. It's essential to recognize that inventory management approaches should be tailored to the strategies, business processes, and values of each company. The effectiveness of inventory management is directly linked to a company's competitiveness and its ability to thrive in the long-term [33].

Inventory management encompasses various activities, including forecasting, calculating safety stock, determining production or procurement quantities, warehouse operations, and more. One crucial aspect of effective inventory management is the forecasting technique. Forecasting is a method that utilizes past data to create informed predictions, allowing us to anticipate the direction of future trends. In supply chain management, forecasts are employed to ensure that the correct product is available at the appropriate location precisely when it is needed.

Effective forecasting plays a pivotal role in helping manufacturers, distributors, and retailers manage their inventory efficiently. It leads to reduced excess stock and prevents unnecessary orders, ultimately contributing to improved profit margins.

4.3.1 Challenges associated with inventory management forecasting

Forecasting and planning for inventory management have garnered substantial attention from the Operational Research (OR) community over the past 50 years. This attention stems from the profound implications it holds for decision-making, spanning both the strategic and operational levels within an organization [34].

Forecasting can significantly enhance an inventory management system by improving decision-making, refining production planning, minimizing overstock, averting stockouts, and optimizing inventory levels [35].

There are numerous types of inventory forecasting methods available. All forecasting models rely on identifying upward and downward trends in consumer demand. Selecting the most accurate forecasting model can be a challenging task, as it requires matching the model to the specific nature of the industry, the size of the company, the production volume, and the location of the company.

The literature discusses numerous challenges associated with forecasting in inventory management, with demand variability and uncertainty being prominent. Fluctuating customer demand due to seasonal trends, market shifts, economic conditions, or unexpected events poses a significant challenge in accurate forecasting and can result in detrimental consequences if not managed effectively.

Inaccurate or incomplete historical data can lead to flawed forecasts. Data that are outdated, unreliable, or insufficient hinder accurate predictions. Therefore, ensuring data quality and availability is crucial for precise forecasting.

In the context of stockouts and overstocking, accurate forecasting plays a pivotal role in maintaining a balanced inventory. Overestimating demand can lead to overstocking, tying up capital and storage space. Conversely, underestimating demand can result in stockouts, leading to customer dissatisfaction and lost sales.

Unforeseen events and changes in economic conditions, including fluctuations in interest rates, inflation, and shifts in customer spending habits, can disrupt demand patterns and further exacerbate the challenge of accurate forecasting.

4.3.2 Benefits of accurate inventory forecasting

Accurate forecasting offers numerous benefits to businesses. Understanding these advantages aids in selecting an appropriate forecasting model that aligns with the company's requirements. Precise forecasting plays a pivotal role in determining the optimal inventory levels, thereby preventing excess stock or stockouts. This optimization effectively reduces carrying costs and maximizes the efficient utilization of working capital.

When forecasts appear accurate, they facilitate improved coordination with suppliers and manufacturers, leading to streamlined production plans and efficient supply chain operations. This, in turn, can result in cost savings and enhanced agility in responding to market changes.

Precision in forecasting ensures a closer alignment between production and demand, which accelerates inventory turnover. Faster turnover rates mean products are sold more swiftly, leading to improved cash flow and reduced holding expenses [34].

5. Conclusions

Supply chain management (SCM) has become a cornerstone of operational efficiency and cost-effectiveness, encompassing material and information flows that influence various facets of production, inventory control, and delivery planning. Supply chain management's integral relationship with inventory management further underscores its importance. Effective supply chain decisions can lead to cost reductions and improved performance. Integrating inventory management into the supply chain context ensures the availability of products at the right time, preventing stockouts and overstocking.

This chapter has explored the critical integration of inventory management within the broader context of SCM. Inventory management holds a pivotal role in enhancing business performance and profitability. The efficient management of inventory levels can significantly boost operational efficiency and optimize financial strategies.

By maintaining optimal stock levels, organizations can enhance operational effectiveness and achieve cost savings. The objectives of inventory management include maintaining working capital, optimizing costs, reducing duplication, and ensuring a steady material flow to support consistent production. The intricate balance of costs, such as carrying, ordering, and stockout costs, plays a pivotal role in determining cost efficiency.

Addressing challenges in inventory management, such as demand forecasting and lead time management, is essential for achieving efficiency. Through an overview of key objectives and techniques, this chapter has contributed to advancing inventory management theory. The discussed techniques, such as ABC inventory classification, Just-In-Time (JIT) management, Economic Order Quantity (EOQ), safety stock, and Vendor-Managed Inventory (VMI), offer distinct advantages for different scenarios. Alongside these techniques, best practices such as accurate forecasting, supplier collaboration, cross-functional collaboration, and regular audits are essential for achieving efficient inventory management.

Moreover, this chapter has highlighted best practices that contribute to effective inventory management. Accurate demand forecasting, supplier collaboration, cross-functional teamwork, and regular audits all play critical roles in maintaining inventory accuracy, enhancing operational efficiency, and aligning strategies with organizational goals.

In conclusion, integrating inventory management within SCM is vital for organizations seeking to thrive in today's interconnected business landscape. By employing appropriate techniques and adopting best practices, businesses can optimize inventory levels, ensure timely customer satisfaction, and achieve long-term financial stability within their supply chain networks. The success of modern businesses depends on their ability to navigate these complexities and effectively manage their inventory to meet customer demands while minimizing costs.

6. Limitations and direction for future work

The chapter presents valuable insights into the realm of inventory management and its pivotal role within supply chains. However, it does exhibit certain limitations that warrant consideration for a comprehensive understanding of the topic. First, while the article provides an overview of various inventory management techniques, it does not delve deeply into the practical implementation challenges and nuances associated with each technique. A more thorough exploration of real-world case studies and examples could elucidate how these techniques perform under diverse industry contexts and shed light on their potential drawbacks and limitations.

Additionally, the article also lacks an in-depth discussion of the trade-offs inherent in each inventory management technique. For instance, while Just-In-Time (JIT) inventory management can minimize carrying costs, it might leave businesses vulnerable to supply disruptions. Addressing these trade-offs and elucidating how industries and companies make decisions based on their unique circumstances would provide a more nuanced understanding of inventory management strategies.

To address these limitations and provide a more well-rounded exploration of inventory management, future research could incorporate practical case studies, recent industry developments, and insights from experts in the field. This would contribute to a deeper understanding of the challenges and opportunities presented by inventory management within modern supply chain contexts.

Author details

Ahmed Esmail Mohamed
University of Southern Denmark, Odense M, Denmark

*Address all correspondence to: ahmed.esmail.mohamed19@gmail.com

References

[1] Mentzer JT et al. Defining supply chain management. Journal of Business Logistics. 2001;**22**(2):1-25

[2] Lambert DM, Cooper MC, Pagh JD. Supply chain management: Implementation issues and research opportunities. The International Journal of Logistics Management. 1998;**9**(2):1-20

[3] Lee HL, Padmanabhan V, Whang S. Information distortion in a supply chain: The bullwhip effect. Management Science. 1997;**43**(4):546-558

[4] Mohamed AE. Moving from the traditional practices to the green era. Research Studies; 4 Mar 2023;**3**(3): 344-361. DOI: 10.55677/ijssers/V03I3Y2023-01. Impact Factor: 5.574

[5] Gumus AT, Guneri AF, Ulengin F. A new methodology for multi-echelon inventory management in stochastic and neuro-fuzzy environments. International Journal of Production Economics. 2010;**128**(1):248-260

[6] Ekanayake N, Joshi N, Thekdi SA. Comparison of single-echelon vs. multi-echelon inventory systems using multi-objective stochastic modelling. International Journal of Logistics Systems and Management. 2016;**23**(2):255-280

[7] Bonney M. Trends in inventory management. International Journal of Production Economics. 1994;**35**(1-3):107-114

[8] Ziukov S. A literature review on models of inventory management under uncertainty. Business Systems & Economics. 2015;**5**(1):26-35

[9] Ali AK. Inventory management in pharmacy practice: A review of literature. Archives of Pharmacy Practice. 2011;**2**(4):151

[10] Hedrick FD, Barnes FC, Davis EW, Whybark DC, Krieger M. Inventory Management. US Small Business Administration; n.d.

[11] Disney SM, Towill DR. A procedure for the optimization of the dynamic response of a vendor managed inventory system. Computers & Industrial Engineering. 2002;**43**(1-2):27-58

[12] Sbai N, Benabbou L, Berrado A. Multi-echelon inventory system selection: Case of distribution systems. International Journal of Supply and Operations Management. 2022;**9**(1):108-125. DOI: 10.22034/ijsom.2021.109031.2138

[13] Cachon GP, Fisher M. Supply chain inventory management and the value of shared information. Management Science. 2000;**46**(8):1032-1048

[14] Lancioni RA, Howard K. Inventory management techniques. International Journal of Physical Distribution & Materials Management. 1978;**8**(8):385-428

[15] Hatefi S, Torabi S, Bagheri P. Multi-criteria ABC inventory classification with mixed quantitative and qualitative criteria. International Journal of Production Research. 2014;**52**(3):776-786

[16] Javadian Kootanaee A, Babu KN, Talari H. Just-in-time manufacturing system: From introduction to implement. Available at SSRN 2253243. Vol. 1. No. 2. Mar 2013. pp. 7-25. ISSN: 2327-8188

[17] Agarwal S. Economic order quantity model: A review. VSRD International

Journal of Mechanical, Civil, Automobile and Production Engineering. 2014;**4**(12):233-236

[18] Silver EA, Pyke DF, Peterson R. Inventory Management and Production Planning and Scheduling. Vol. 3. New York: Wiley; 1998

[19] Rădăşanu AC. Inventory management, service level and safety stock. Journal of Public Administration, Finance and Law. 2016;**2**(9):145-153

[20] Sari K. Exploring the benefits of vendor managed inventory. International Journal of Physical Distribution & Logistics Management. 2007;**37**(7):529-545

[21] Tian X, Wang H, Erjiang E. Forecasting intermittent demand for inventory management by retailers: A new approach. Journal of Retailing and Consumer Services. 2021;**62**:102662

[22] Bon AT, Leng CY. The fundamental on demand forecasting in inventory management. Australian Journal of Basic and Applied Sciences. 2009;**3**(4):3937-3943

[23] Mohamed AE. An AHP framework to evaluate barriers and potential tensions to green supply chain management in the food and beverage industry. European Journal of Business and Management. 2021;**13**(6). 2222-1905 (Paper). ISSN: 2222-2839 (Online). DOI: 10.7176/EJBM/13-6-01

[24] Mohamed AE. Mitigating the bullwhip effect and enhancing supply chain performance through demand information sharing: An ARENA simulation study. Journal of Economics and Sustainable Development. 2023;**14**(14). 2222-1700 (Paper). ISSN: 2222-2855 (Online). DOI: 10.7176/JESD/14-14-07

[25] Yao Y, Evers PT, Dresner ME. Supply chain integration in vendor-managed inventory. Decision Support Systems. 2007;**43**(2):663-674

[26] Omar IA, Jayaraman R, Salah K, Debe M, Omar M. Enhancing vendor managed inventory supply chain operations using blockchain smart contracts. IEEE Access. 2020;**8**:182704-182719

[27] Waller M, Johnson ME, Davis T. Vendor-managed inventory in the retail supply chain. Journal of Business Logistics. 1999;**20**(1):183

[28] Achabal DD, McIntyre SH, Smith SA, Kalyanam K. A decision support system for vendor managed inventory. Journal of Retailing. 2000;**76**(4):430-454

[29] Southard PB, Swenseth SR. Evaluating vendor-managed inventory (VMI) in non-traditional environments using simulation. International Journal of Production Economics. 2008;**116**(2):275-287

[30] Groenvelt H. The just-in-time system, in handbooks in operations research and management science. In: Graves SC, Rinnooy Kan AHG, Zipkin PH, editors. Logistics of Production and Inventory. Vol. 4. North Holland, Amsterdam: Elsevier Science; 1993. pp. 629-670

[31] Wilson JM. Henry Ford's just-in-time system. International Journal of Operations & Production Management. 1995;**15**(12):59-75

[32] Kros JF, Falasca M, Nadler SS. Impact of just-in-time inventory systems on OEM suppliers. Industrial Management & Data Systems. 2006;**106**(2):224-241

[33] Nugroho RE, Resodiharjo M. Inventory management analysis by optimizing the forcasting methods (case study at PT XYZ Indonesia). Dinasti International Journal of Management Science. 2021;**2**(3):435-455

[34] Syntetos AA, Boylan JE, Disney SM. Forecasting for inventory planning: A 50-year review. Journal of the Operational Research Society. 2009;**60**(sup1):S149-S160

[35] Chowdhury MH, Ahmed T, Rahman MB, Islam AS. A smart inventory management system with forecasting technique applied to efficiently handle industrial asset. American Journal of Innovation in Science and Engineering. 2023;**2**(2):1-5

Chapter 4

Emerging Inventory Planning

Andrés Felipe Santos Hernández and Natalia Camacho Franco

Abstract

In the last decade, the industry has been driven by the fifth industrial revolution, but it has also had to face unforeseen challenges in the economic, political, and international conflict arenas. These factors have generated difficulties that impact the efficient operation of global supply chains. In response to this uncertainty, many organizations have chosen to stockpile large quantities of inventory, increasing working capital while reducing their Return on Assets (ROA). This crisis has brought about swift changes in markets, with shifts in purchasing behavior and increasing competition in terms of prices and offers. Organizations need to use knowledge and technology more intelligently in this dynamic context to design strategies that enhance inventory planning and management. This chapter will explore the significance of logistical conditions, planning, and several strategies to achieve a responsive inventory in an ever-evolving market, where inefficiency and poor service are no longer tolerated.

Keywords: inventories, optimization, planning, emerging markets, aggregate planning, supply strategies, algorithms

1. Introduction

Over the past decade, the global economy has been characterized by unpredictable behavior, fueled by a multi-crisis driven by the post-pandemic era, the Ukraine conflict, and now the turmoil in the Arabian region. These situations have brought to light subsequent difficulties such as an energy crisis, inflation, food insecurity, and security concerns. These effects translate into present risks in the medium term, in which economic slowdown, high living costs, societal divisions, and misinformation through digital channels can destabilize the performance of global supply chains [1, 2].

The interconnectivity of these supply chains is determined by economies of scale and the flows represented in geography. Countries with limited maritime connections face specific constraints as they must frequently engage in container or loose cargo transshipments. This has led to temporal dispersion, less reliable connections, and, consequently, a lower logistical performance, resulting in high operating costs for various organizations [1].

In light of the recent risks and disruptions in supply chains, it becomes important not only to quantify costs but also to focus on planning the length, the network's specific characteristics, and the sector's resilience in all logistical processes. To achieve the improved logistical performance, it is crucial to revise certain organizational policies. Additionally, the academic community plays a constructive role in enhancing tools, technology, and, consequently, decision-making to gain an ever-increasing competitive advantage in economic sectors [2].

In addition to the risks experienced in the performance of supply chains, competition and the quality of logistical services are crucial factors for organizations. According to the World Bank's Logistics Performance Index, which covers 139 countries, these elements significantly impact overall performance.

This report assesses six variables, including customs, infrastructure, international shipments, logistics competence, traceability, and shipping times. **Figure 1** displays the scores of key representative countries for each variable. For instance, in logistics competence, Singapore achieved the highest score with 4.4, Mexico fell around the average with a score of 3.0, and Somalia had the lowest score at 1.8 out of 5.0.

For the countries under study, the objective of this research is to compare their logistical performance, identify key challenges in their international logistics, gain visibility into global trends, and discover the improvement opportunities to enhance productivity in supply chains [3].

In light of this study, 75 out of the 139 countries do not exhibit high logistics competence, meaning their indicators are close to 3.0. Countries such as Benin, Mexico, Panama, Rwanda, Chile, Colombia, Hungary, and Uruguay have made efforts to enhance their logistical conditions, achieving modest progress. However, when it comes to knowledge and the application of models in logistics, they still have room for improvement [3].

Upon reviewing the data provided by the World Bank since 2007, the Colombian government has shown a commitment to enhancing the logistical conditions within the country. Through the National Planning Department (DNP), national surveys have been conducted to identify the characteristics and improvement opportunities that entrepreneurs can explore within their supply chains.

The latest edition of this survey was conducted in 2020, with a probabilistic sample of 3383 companies. It evaluated five key areas:

- Logistical Performance: This assesses operational costs, service quality, and the utilization of technology in logistics.

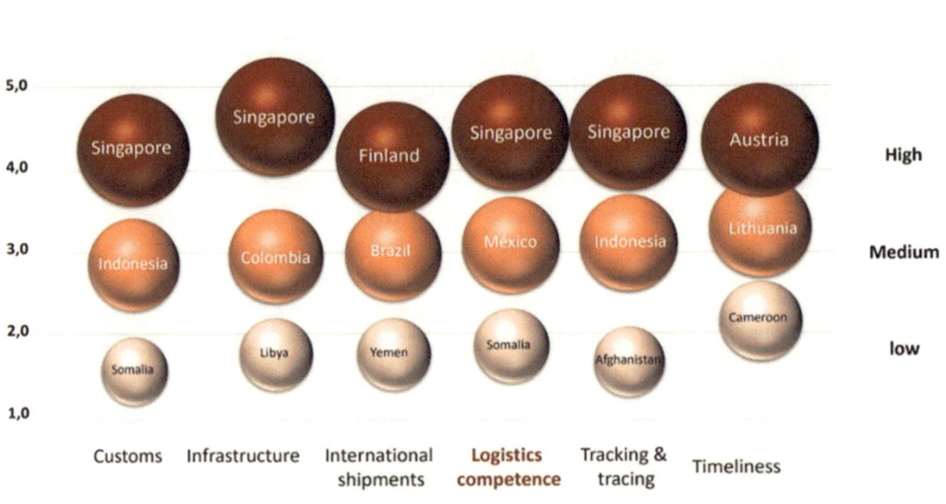

Figure 1.
Logistics performance index variables 2023.

- Outsourcing of Logistical Services: This determines the extent to which companies contract out logistical services (involving 3, 4, or 5 logistical parties) within their logistics processes, aiming for efficiency in time and costs.

- International Logistics: It evaluates foreign trade processes, including cost, customs regulatory compliance, and ease of international trade.

- Regional Logistics: This aims to highlight challenges and issues in different regions, assessing the availability of logistical services in these areas.

The Future of Logistics in the Country: The objective here is to identify challenges in operations aligned with the needs, such as road, river, rail, port, and airport infrastructure, that the government can address through the design and implementation of public policies [4].

One of the most relevant indicators in this study is logistical performance. Colombia's economic openness and tariff negotiations with other countries have turned logistical costs into a key variable for organizational competitiveness in this country. Measuring the cost of logistics at the national level has enabled the government to design more effective policies to enhance the logistical conditions that various companies require for managing their supply chains at a competitive level [5].

In the case of this country, logistical costs are measured as a percentage of its sales. For the year 2020, the indicator had a value of 12.6% of sales, with a target of 13.3%, as shown in **Figure 2**. In this report, the highest costs were incurred in warehousing, accounting for 47%, and transportation, accounting for 35%. Due to the pandemic, consumers adapted to the digital market, grew accustomed to swift response times, and no longer wanted to wait. They assumed new roles within their families, transforming business and logistical practices for companies. This created the need for more competitive resources in the relevant areas. However, stock levels increased to keep up with this consumption pace. Unfortunately, organizations lacked a suitable transactional system to keep up with demand, resulting in high costs for warehousing and inventory [4].

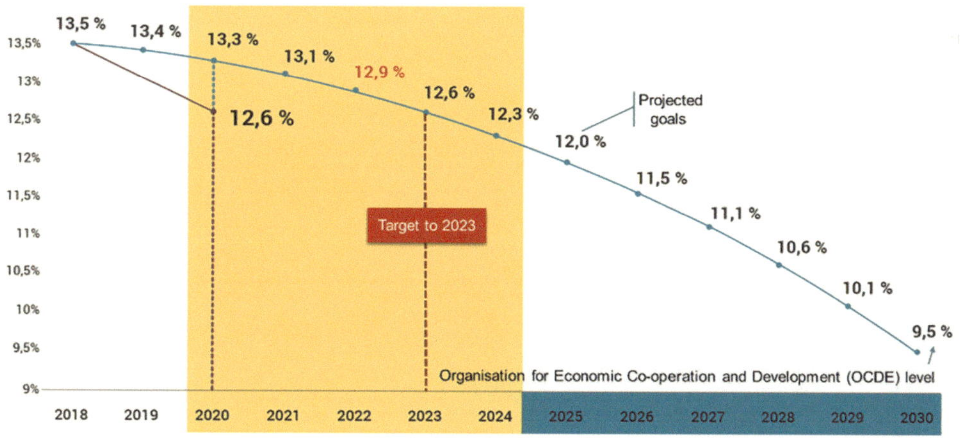

Figure 2.
Logistics cost; percentage of sales in Colombia.

Globally, high inventory levels primarily affect the retail sector. Currently, this group is experiencing a decline in its sales. Among several challenges, companies are dealing with high inflation, supply chain issues, evolving consumer behaviors, labor shortages, global conflicts, climate change, and the pressing need for improved technology to meet the demands of a more demanding consumer base [6].

By the statement made by Juan Germán Osorio, Consumer Leader Partner at Deloitte Americas [7]:

"In the world, seven out of ten executives in the food and beverage, household goods, personal care, or clothing industries claim that their job is more stressful today than it was 5 years ago. This is attributed to factors such as record-high inflation, supply chain issues, evolving consumer behaviors, labor shortages, global conflicts, and climate change".

These words are derived from a report released by Deloitte in 2023, titled "Consumer Products Industry Outlook to 2023". This study conducted surveys with 150 global consumer product companies in the food and beverage, household goods, personal care, and clothing industries. The data sources are multinational industries with annual revenues exceeding $500 million. As mentioned earlier, Deloitte has confirmed that organizations are facing new challenges, as indicated in **Figure 3**. The market landscape has transformed into a dynamic scenario where organizational planning must be more attuned to demographic, political, environmental, technological, and cultural changes.

Among the surveyed group, there are organizations with profitable growth that are implementing the following strategies [8]:

- 93% adapt to the changing consumer and align with their dynamics.

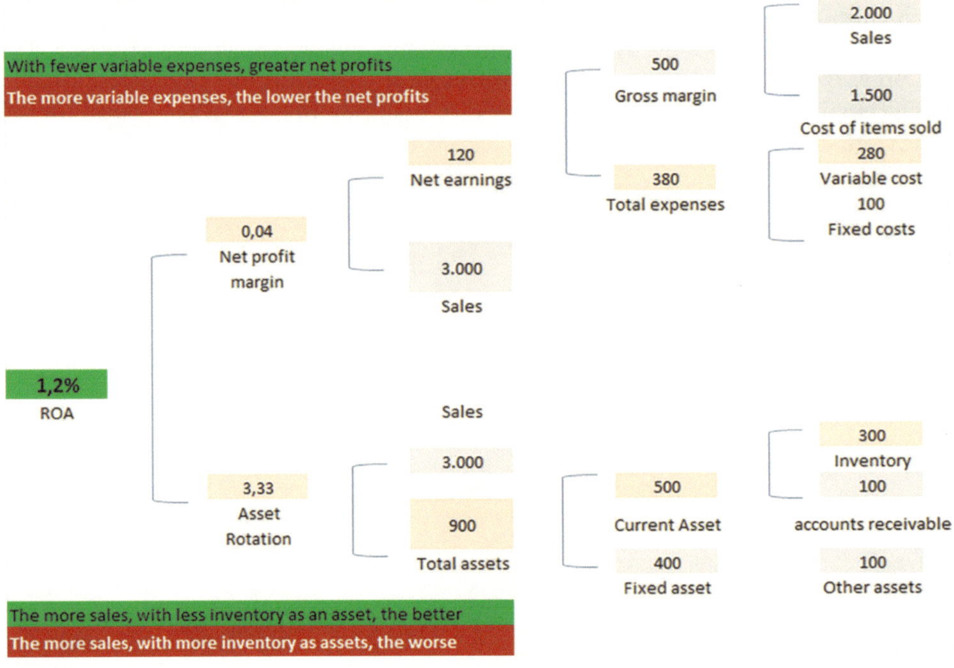

Figure 3.
Inventory's effect on ROA.

- 85% believe that the current environment provides an opportunity to increase market share.
- 68% view vertical integration as a strategic approach.
- 90% are investing in technology to have real-time information throughout their supply chains.
- 97% consider environmental issues a priority for organizations.

Of all these perceptions, technology is one of the most significant. In the survey, nine out of ten companies are investing in enhancing supply chain management and operational excellence. They are also investing in big data to share information with their suppliers and distribution channels, to coordinate and optimize logistics. Ultimately, all the strategies and tactics managed along this chain will be reflected in the price, determining whether or not a competitive advantage is achieved [8].

Organizations cannot move forward independently; they need to have a partner and invest in Artificial Intelligence (AI). Sales channels and the multitude of Stock Keeping Units (SKUs) can no longer be sustained solely by the workforce. They require assistance with predictive analytics and mathematical models that can optimize inventory and provide responses to customers, almost in real time [9].

In the face of an economic slowdown in 2023, organizations are beginning to encounter problems and more challenges, such as a lack of investment, limited working capital, and an increase in competitive rivalry [10]. This, coupled with a lack of visibility and synchronization from the sourcing of materials to the storage of finished products in retail channels, has led to the accumulation of high inventory levels in various supply networks, compromising profitability on assets (ROA) and return on investment (ROI). This, in turn, reduces inventory turnover and, therefore, the company's profitability, especially in times of crisis.

2. Inventory as a strategic asset

Inventory, while basic to many, is an essential and strategic component in the operation of organizations. As a significant entity among current assets, inventories represent an organization's ability to produce, store, and market goods and services. Likewise, their proper management has a significant impact on profitability, service levels, and the competitiveness of the business.

As Coyle mentions, "Inventory as an asset on the balance sheet and as a variable expense on the income statement has gained significant importance as organizations seek more effective ways to manage their assets and working capital" [11].

The ability of inventories to generate economic value through their sale or use in the production of goods or services for marketing makes them a primary asset for organizations seeking to improve their profitability. Managing variables related to their management is key to achieving cost savings in operations [12, 13].

Among the most important current assets on the balance sheet are cash, accounts receivable, and inventories, with the latter being the most valuable. However, the lack of assessment and calculation of inventory for each item can lead to unnecessary expenses in the income statement, posing a challenge for organizations [12]. The management of this asset determines the return on investment (ROI), and as Coyle and other authors mention, "reducing inventories leads to a short-term increase in ROI as assets decrease and available working capital increases, while an increase in inventories results in an increase in assets, causing a decrease in the same indicator and available working capital" [11]. In this sense, considering inventory as an asset, one must monitor the return on assets (ROA).

Similarly, the value of inventory influences the return on assets (ROA) of an organization when considered as an asset. **Figure 3** illustrates how the inventory level affects each phase of the financial relationship until reaching this indicator.

Considering this relationship, if a company aims to increase its return on assets (ROA), it should redefine its procurement policy to control the quantities and frequencies of procurement. Similarly, it should review operational costs and their utilization, such as transportation frequency, storage costs, inventory, and maintenance. However, it is essential to ensure that the percentage variation of these variables is in line with the capacity and nature of the business to avoid compromising the quality of the delivery service.

Additionally, to maintain this level of service, it is important to consider the existing relationship between transportation and inventory. If there is a decrease in inventory, it will lead to an increase in transportation costs, and vice versa, all to fulfill the sales promise. For this reason, when redefining the procurement policy, it is crucial to always seek a balance or trade-off between these variables. The company's policy will determine which strategy benefits the company more: reducing transportation costs or inventory costs while maintaining a high fill rate[1].

Another one of the main and most common challenges faced by some organizations is the proper management of inventories to the demand encountered. Well-managed inventory can help improve operational efficiency, reduce costs, and meet customer demand. However, poorly managed inventory can lead to unnecessary costs and product shortages. At this point, companies must present a solid planning strategy because, depending on the market type and corporate strategy, it is necessary to have enough inventory to meet demand and avoid losses due to stock-outs. However, it is not advisable to have too much inventory, as it poses a significant risk of increased operational costs [14].

Achieving the balance between maintaining sufficient stock without excess has a significant impact on various areas, and some of the main benefits that can be achieved include [15, 16]:

Cost Optimization: Avoiding storage costs due to excess inventory while also preventing additional costs from production delays or penalties for inventory shortages.

Customer Satisfaction: Maintaining sufficient stock ensures that the company can meet customer demand promptly, enhancing customer satisfaction.

Risk Minimization: Excess inventory increases the risk of product deterioration, which can result in financial losses. On the other hand, inventory shortages can reduce the level of service.

Adaptation to Demand Changes: A proper inventory balance allows the company to quickly adjust to changes in market demand. If demand suddenly increases, the company can meet it without delays or sales losses. If demand decreases, the company would not be burdened with excess unsold inventory.

---

[1] In logistics, this indicator is used to measure the efficiency of the delivery service to customers.

Cash Flow Improvement: Excess inventory negatively impacts available cash flow, while insufficient inventory can lead to sales losses, also affecting cash flow.

On the other hand, the inventory value represents a significant influence in the study of the financial health of the company, with which some factors can be analyzed, such as [17, 18]:

- Solvency and Liquidity: The inventory level is related to financial obligations, where an appropriate level of inventory indicates that the company can meet its obligations and debts.

- Profitability: Efficient inventory management can enhance profitability by reducing purchasing and storage costs.

- Operational Efficiency: Excessive inventory suggests inefficiency and additional costs, while low inventory indicates service level issues.

In conclusion, inventory becomes a strategic asset that improves a company's profitability and success. Proper management has a significant impact on the financial health of the organization and its ability to meet market demand. Next, we will present some methods and practices for achieving proper and optimal inventory planning.

3. Aggregate inventory horizon

An effective methodology for comprehensive planning across all three inventory levels is the aggregate plan. This technique allows for the determination of optimal levels of capacity, production, and inventory over a specific period to meet demand and maximize profits.

This planning over the horizon emerges as an essential strategy in the management of operations and assets in the supply chain of any organization. Its primary goal is to achieve an optimal balance between demand, production capacity, and inventory turnover, thereby facilitating effective resource planning and decision-making [19].

This planning focuses on medium-term decision-making, aiming to optimize plant capacity utilization and the strategic use of inventories. It should not be regarded as a predictive tool because it involves decisions at the level of aggregated units and families, rather than specific decisions for individual stock-keeping units (SKU) [20].

To achieve effectiveness and reliability in the model, data from each of the operations in the process being evaluated are required. This requirement poses one of the greatest challenges for some companies that either have not collected the data or choose not to share it with their suppliers or customers. Without these data, which serve as a fundamental input in planning, whether they are related to costs or constraint conditions, it will be difficult to achieve the expected outcome of this planning.

This planning model provides a valuable tool for the production management and control process, as well as for strategic control. It facilitates effective communication among different departments and areas to achieve the company's strategic objectives. However, some of the challenges that arise in the application of this method include:

- Market Demand Variability: Companies must understand how to reduce or mitigate the effects of market demand variability through flexibility and adaptability.

- Collaboration with suppliers and distributors is essential for achieving successful aggregate planning. Establishing strong relationships and sharing information is key to this collaboration.

By the suggestions resulting from the model, the supply chain management should be aware of which tactics to employ to align the strategy within the organization. Defining inventory policies for raw materials, work-in-progress, and finished products in terms of inventory days is crucial for the tactical level, meaning leveraging inventories to respond to the market when capacity is limited. Flexible scheduling is related to using time as a lever, where capacity remains constant, but the number of hours worked is adjusted in response to changes in demand. Finally, capacity acts as a lever that allows for the synchronization of production based on demand fluctuations, involving capacity adjustments through machinery modifications or hiring and firing employees [19].

3.1 Study case

Steel Wire, a company in the steel industry, is facing the challenge of optimizing its production and inventory planning due to inadequate plant capacity to meet the growing demand for its primary product, the 80-kg Link Mesh. To address this situation, a specific aggregate planning model is being developed for this product.

This aggregate planning model will be a valuable tool to help Steel Wire balance demand and production capacity, enabling it to make strategic decisions regarding resource allocation and inventory management. Additionally, the company can consider options such as production flexibility, collaboration with suppliers and distributors, and inventory policy optimization to ensure an effective approach in this situation.

The production of the 80-kg Link Mesh involves three fundamental stages: wire drawing, galvanizing, and transformation. The process begins with the supply of material, which consists of an SAE 1006 steel wire rod with an initial diameter of 5.5 mm. This type of wire rod is purchased in 2-ton coils supplied by Paz del Río, a steel company located in Boyacá, Colombia.

These wire rod coils are transported from the raw materials warehouse to the two wire drawing machines of each drawing bench using forklifts. During the wire drawing phase, an operator is responsible for supervising and operating two semi-automatic wire drawing machines: The Kock 3 and the Kock 4.

The wire drawing operation involves the plastic deformation of steel through continuous tensile stresses, where the wire passes through a set of dies [21]. For this particular product, a diameter of 3.28 mm is required. The resulting wire is wound into coils, which can weigh up to 900 kg each. These coils are then transported by the forklift to the hot-dip galvanizing area. On each trip, the vehicle can transport up to four coils at a time.

The galvanizing process has a capacity of 50 wires, of which five are programmed to supply the mesh production line. This process begins with the degreasing stage, where organic contaminants are removed, pickling is continued to eliminate the oxide scale on the wire's surface, and then, fluxing is done. Fluxing uses a saline solution of ammonium chloride and zinc to remove any remaining oxide, providing a protective layer to the steel. Finally, the galvanizing process is carried out by immersing the 50 wire lines in a bath of molten zinc, the conditions of which comply with the specifications of the American standard ASTM B6. This process requires a minimum of 98% pure zinc and a heating temperature within the range of 435–455°C [22, 23].

Once the material has passed through the galvanizing process, its diameter is adjusted to 3.30 mm as the wire is coated with a 0.2-mm-thick layer of zinc. After zinc immersion, the wire is wound into 900 kg coils, which are then transported to the production area of the link mesh. In this section, there are 8 Waffios machines programmed to transform the wire into rolls of link mesh, each weighing 80 kg.

Each machine requires two coils to interlace the link mesh, as shown in **Figure 4**. The operator of each machine is responsible for threading the wires and adjusting the machine's settings, which operate semi-automatically. Once the manufacturing process of a roll is completed, the operator places cardboard and labels at the ends, wraps the roll with stretch paper, and stacks it on a pallet (**Figure 5**).

The data associated with this process are indicated in **Table 1**.

The aggregate plan is a model designed to maximize profitability over a specific time horizon within a given process. In the case of an organization in the steel industry, its primary focus is to provide maximum value to customers by committing to meet all projected demand, even if this entails certain delays. The key to maximizing profitability lies in operational efficiency, with the additional consideration of pricing and promotion policies. By effectively minimizing operating costs, a significant contribution is made to profit maximization, always in line with the company's strategy.

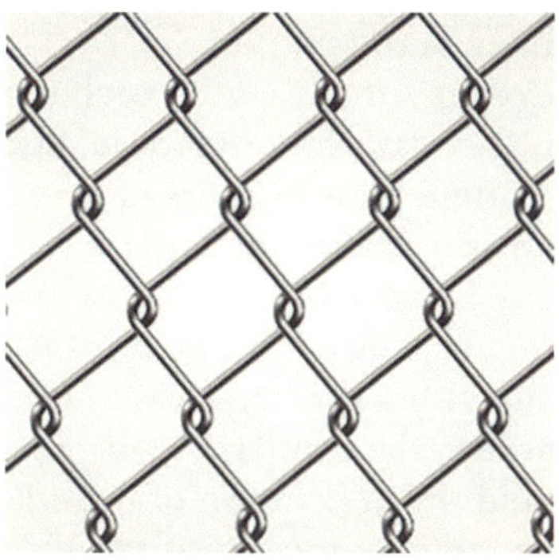

Figure 4.
Chain link mesh detail.

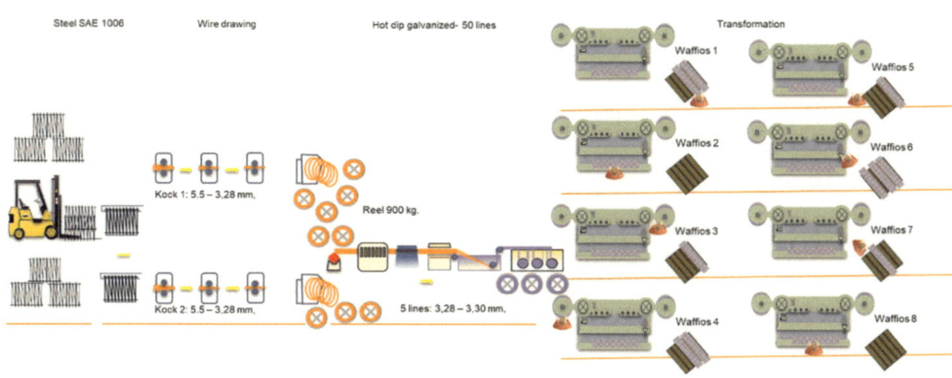

Figure 5.
Chain link mesh process.

Operation	Equipment	Production speed (Ton/Shift)	Efficiency (%)	Real production (Ton/shift)
Drawing	Kock 3	8	66	5.3
	Kock 4	8	66	5.3
Hot dip galvanized	5 lines	14	70	9.8
	wafios 1	1.28	75	0.96
	wafios 2	1.28	75	0.96
	wafios 3	1.28	75	0.96
	wafios 4	1.28	75	0.96
Transformation	wafios 5	1.28	75	0.96
	wafios 6	1.28	75	0.96
	wafios 7	1.28	75	0.96
	wafios 8	1.28	75	0.96

Table 1.
Standard data of the chain link mesh process x 80 kg.

In this regard, this company aims to optimize the planning of the 80-kg Link Mesh, one of its high-demand products in the market.

The process operates for 25 days a month, with each worker putting in 8 hours per day in regular time. However, they can work overtime, but according to the country's legislation, they cannot exceed more than 15 hours of overtime per month. The production operation's capacity is primarily determined by the total hours of work employed. Therefore, the machine capacity is not a limiting factor for the production operation.

The monthly subcontracting capacity is set at 50 tons, while the inventory has a capacity of 80 tons per month, with a shortage limit of no more than 0.5 tons per month. At the end of the semester, the company will ensure the maintenance of 5 days of available inventory and avoid the generation of orders in a state of shortage.

The design of this plan aims to maximize profitability over the semester while ensuring an adequate level of service for the market. This will be achieved by assuming that prices will remain constant during this period and that the task of reducing costs will be determined by the model, allowing us to optimize profitability. The forecast for the upcoming months is illustrated in **Table 2**.

At the end of June, the following figures were recorded: 8 workers in the Link Mesh production line, 2 tons of work-in-progress in the wire drawing stage, 4 tons of work-in-progress in the galvanizing stage, 7 tons of finished product, and 6.2 tons in a shortage state. The costs associated with this process are detailed in **Table 3**.

To calculate the cost per kilogram of operations like wire drawing, galvanizing, and transformation, the actual production per hour (**Table 1**) is divided by the hourly cost for each area or line (**Table 2**). The result is presented in **Table 4**. In the case of the mesh area, the cost of employed workers is added to this calculation, which is subject to market demand and the established inventory policy.

Additionally, the organization needs to establish its work-in-progress inventory policy. The number of inventory days defines the availability period for this inventory. For example, if there is a 5-day work-in-progress inventory, it means there is availability to cover 5 days of operation.

Month	Demand (Kg.)
July	320,000
August	290,000
September	260,000
October	210,000
November	240,000
December	300,000

Table 2.
Estimated demand for chain-link mesh.

Process	Cost	Unity
Material	4,500	$ / kg
Inventory holding cost	500	$ / kg/ month
Cost of stock-outs	1,500	$ / kg/ month
Hiring + training cost	1,150,000	$ / worker
Layoff cost	1,400,000	$ / worker
Hours MOD transformation (1 Operator - 1 Machine)	120	Kg. / hour
Cost of overtime MOD transformation	15,500	$ / hour
Drawing cost + operator	9,500	$ / hour
Galvanizing cost + operator	11,500	$ / hour
Transformation cost (Only machine)	6,500	$ / hour
Outsourcing cost	15,000	$ / kg

Table 3.
Costs associated with the production process of chain link mesh.

Operation cost	Standard production (Kg. / hour)	Cost ($ / hour)	Cost/kg.
Drawing	1,320	9,500	7.20
Galvanizing	1,225	11,500	9.39
Transformation	960	6,500	6.77

Table 4.
Costs per kg. Of production process of chain link mesh.

The purpose of this policy is to control working capital and ensure a buffer throughout the entire process. This policy is defined for all three operations, as well as for finished product inventory and the allowed delay in each month, as shown in **Table 5**.

3.2 Case solution

In order to find a reliable and time-efficient solution, the case study is solved using an algorithm implemented in the Python programming language. The resolution process consists of several steps:

Product in process		Finish product - 5 days of inventory)	Backlog (Kg.)
Drawing - 3 days of inventory (Kg.)	**Galvanizing - 4 days of inventory**		
32,000	42,667	53,333	500
29,000	38,667	48,333	500
26,000	34,667	43,333	500
21,000	28,000	35,000	500
24,000	32,000	40,000	500
30,000	40,000	50,000	0

Table 5.
Inventory policies.

3.2.1 First steep

In this phase, the "pulp", "pandas", and "numpy" libraries are imported to leverage the linear and integer programming problem-solving capabilities provided by "pulp". Additionally, the data manipulation and numerical operation functionalities provided by "pandas" and "numpy" are used.

```
%pip install pulp
from pulp import*
import pandas as pd
import numpy as np
```

3.2.2 Second steep

At this point, we start defining a linear programming model. This begins by creating a list that maps the defined decision variables, whose values will be determined by the aggregate plan. Twelve variables are defined over a range of six periods (**Table 6**).

Next, a cost matrix is created to represent the costs associated with various variables and activities. Demand is defined for each period, and initial variables such as workforce, stock-out, inventory, and others are initialized. Additionally, four lists are created that contain the monthly inventory values for drawing, galvanizing, finished product, and stock-out.

```
explica={1:'hire',2:'fire',3:'workforce',4:'Overtime',5:'prod_draw_out', 6:'prod_GVZ' ,7:'prod_mesh',8:'inv_draw_out' ,
        9:'inv_GVZ' ,10:'inv_fp', 11:'stock_out' , 12:'outsourcing'}

 Periods=6
 vd=12
 costs = np.array([[500],[1500],[15000],[1150000],[1400000],[6500],[15500],[4507],[9.4],[6.8],[500],[500]])
 demand=np.array([320000,290000,260000,210000,240000,300000])
 w0=8 #fuerza de trabajo inicial
 stockout0=6200
 inv_draw_out0=2000
 inv_GVZ0=4000
 inv_fp0=7000
 days_inv_draw_out= np.array([32000,29000,26000,21000,24000,30000])
 days_inv_GVZ= np.array([42667,38667,34667,28000,32000,40000])
 days_inv_fp= np.array([53333,48333,43333,35000,40000,50000])
 stock_out= np.array([500,500,500,500,500,0])
```

Variable	Description
Hire $_t$	Number of workers hired in month t.
Fire $_t$	Number of workers laid off in month t.
Workforce $_t$	Number of active operators in month t.
Overtime	Number of overtime hours in month t
Prod_draw_out $_t$	Number of kilograms produced in the drawing mill in month t.
Prod_GVZ $_t$	Number of kilograms produced on the galvanizing line in month t.
Prod_mesh $_t$	Number of kilograms produced on the processing line in month t.
Inv_draw_out $_t$	Inventory of drawn material in kilograms in month t.
Inv_GVZ $_t$	Inventory of galvanized material in kilograms in month t.
Inv_FP $_t$	Inventory of 80-kg chain-link mesh in kilograms in month t.
Stock_out $_t$	Kilograms of product overdue for delivery to the customer in month t.
Outsourcing $_t$	Number of kilograms produced by subcontracting in month t.

Table 6.
Model decision variables.

3.2.3 Third steep

In this stage, the linear programming model is created by defining the objective function of the problem, which aims to minimize the total cost incurred during the planning horizon. The total cost is defined by the following components:

- Regular labor time cost

$$\sum_{t=1}^{6} Workforce_t \cdot 6500 \tag{1}$$

- Overtime labor cost

$$\sum_{t=1}^{6} Overtime_t \cdot 15500 \tag{2}$$

- Hiring and firing costs

$$\sum_{t=1}^{6} Hire_t \cdot 1150000 + \sum_{t=1}^{6} Fire_t \cdot 1400000+ \tag{3}$$

- Inventory and stock-out costs

$$\sum_{t=1}^{6} Inv\,FP_t \cdot 500 + \sum_{t=1}^{6} Inv\,draw\,out_t \cdot 500 + \sum_{t=1}^{6} Inf\;GVZ_t \cdot 500 + \sum_{t=1}^{6} Stock\,out_t \cdot 1500 \tag{4}$$

- Outsourcing cost

$$\sum_{t=1}^{6} Outsourcing_t \cdot 15000 \tag{5}$$

- Production cost

$$\sum_{t=1}^{6} \text{Prod}\, draw\, out_t \cdot 4507 + \sum_{t=1}^{6} \text{Prod}\, GVZ_t \cdot 9.4 + \sum_{t=1}^{6} \text{Prod}\, mesh_t \cdot 6.8 \tag{6}$$

The sum of the mentioned costs results in the total cost, thus defining the following objective function denoted by Z:

$$\begin{aligned} Min\, Z = {} & \sum_{t=1}^{6} Workforce_t \cdot 6500 + \sum_{t=1}^{6} Overtime_t \cdot 15500 + \sum_{t=1}^{6} Hire_t \cdot 1150000 + \sum_{t=1}^{6} Fire_t \cdot 1400000 \\ & + \sum_{t=1}^{6} Inv\, FP_t \cdot 500 + \sum_{t=1}^{6} Inv\, draw\, out_t \cdot 500 + \sum_{t=1}^{6} Inv\, GVZ_t \cdot 500 \\ & + \sum_{t=1}^{6} Stock\, out_t \cdot 1500 + \sum_{t=1}^{6} Oursourcing_t \cdot 15000 + \sum_{t=1}^{6} \text{Prod}\, draw\, out_t \cdot 4507 \\ & + \sum_{t=1}^{6} \text{Prod}\, GVZ_t \cdot 9.4 + \sum_{t=1}^{6} \text{Prod}\, mesh_t \cdot 6.8 \end{aligned} \tag{7}$$

The “LpProblem” object from the “pulp” library is used to define the minimization problem. A list is created to store variable names composed of the numeric value of the period (from 1 to 6) and the variable number (from 1 to 12), along with their respective descriptions.

Next, a matrix is defined to organize the decision variables. The objective function is calculated as the weighted sum of the decision variables and their corresponding costs. The model is set up to minimize the objective function.

```
v_names=[str(i)+ explica[i]+str(j)for j in range(1,Periods+1)for i in range(1,vd+1)]
v_names.sort()
v_names
```

```
vdd=LpVariable.matrix('',v_names,cat='Integer',lowBound=0)
print(vdd)
allocation = np.array(vdd).reshape(vd,Periods)
fo= lpSum(allocation*costs)
model += fo
print(fo)
allocation
```

DOI: http://dx.doi.org/10.5772/intechopen.114025

3.2.4 Fourth step

In this step, constraints are introduced into the linear programming model:

3.2.4.1 Labor force restriction

It is established that the workforce must not exceed the number of available machines, as each machine should be operated by a single worker, and it ensures that the sum of the workforce from the previous month, minus the workforce allocated in the current month, plus the number of workers hired and minus the number of workers fired in the same month, equals zero.

$$Workforce_t = Workforce_{t-1} + Hire_t - Fire_t \tag{8}$$

```
for i in range(0, Periods):
  print(lpSum(allocation[5][i])<=8)
  model+=lpSum(allocation[5][i])<=8

model+=lpSum(w0)+lpSum(allocation[3][0])-lpSum(allocation[4][0])-lpSum(allocation[5][0])==0
print(lpSum(w0)+lpSum(allocation[3][0])-lpSum(allocation[4][0])-lpSum(allocation[5][0])==0)
model

for mes in range(1,Periods):
  print(lpSum(allocation[5][mes-1])+lpSum(allocation[3][mes])-lpSum(allocation[4][mes])-lpSum(allocation[5][mes])==0)
  model+=lpSum(allocation[5][mes-1])+lpSum(allocation[3][mes])-lpSum(allocation[4][mes])-lpSum(allocation[5][mes])==0
model
```

3.2.4.2 Constraint on overtime hours

A loop is created that iterates through each period, setting the overtime hours to be less than or equal to 15, as stipulated in the case study.

$$Overtime_t \leq 15 \cdot Workforce_t \tag{9}$$

```
for i in range(0,Periods):
  print(-lpSum(allocation[6][i])+ lpSum(15*allocation[5][i])>=0)
  model += -lpSum(allocation[6][i])+lpSum(15*allocation[5][i])>=0
```

3.2.4.3 Drawing, galvanizing, and transformation capacity constraint

A loop is set up in which, in each iteration, it is determined that the amount of material produced in each process must be less than or equal to the available monthly capacity of both workers and machines.

$$prod\,draw\,out_t \leq 3\,turn \cdot 8\,\frac{hours}{turn} \cdot 25\,\frac{days}{month} \cdot 1320\,\frac{Kg}{hour} \tag{10}$$

$$prod\,GVZ_t \leq 3\,turn \cdot 8\,\frac{hours}{turn} \cdot 25\,\frac{days}{month} \cdot 1225\,\frac{Kg}{hour} \tag{11}$$

$$prod\,mesh_t \leq 3\,turn \cdot 8\,\frac{hours}{turn} \cdot 25\,\frac{days}{month} \cdot 960\,\frac{kg}{hour} \cdot Workforce_t + 960\,\frac{kg}{hour} \cdot overtime_t \tag{12}$$

```
for mes in range(0,Periods):
  print(lpSum(3*8*25*960*allocation[5][mes])+lpSum(960*allocation[6][mes])-lpSum(allocation[9][mes])>=0)
  model+=lpSum(3*8*25*960*allocation[5][mes])+lpSum(960*allocation[6][mes])-lpSum(allocation[9][mes])>=0
model
```

```
for mes in range(0,Periods):
  print(lpSum(3*8*25*1320)- lpSum(allocation[7][mes])>=0)
  model += lpSum(3*8*25*1320)- lpSum(allocation[7][mes])>=0
model
```

```
for mes in range(0,Periods):
  print(-lpSum(3*8*25*1225)+ lpSum(allocation[8][mes])<=0)
  model += -lpSum(3*8*25*1225)+ lpSum(allocation[8][mes])<=0
model
```

3.2.4.4 Drawing, galvanizing, and finished product inventory balance constraint

To balance inventory and production, a loop is created in which, in each period, it is ensured that the inventory of the corresponding process in a month "i" is equal to the inventory of finished products and the stock-out from the previous month, plus the total production and outsourcing of month "i" minus the stock-out from the previous month and the demand of month "i".

$$Inv\,GVZ_t = Inv\,GVZ_{t-1} + \text{Prod}\,GVZ_t - \text{Prod}\,mesh_t \tag{13}$$

$$Inv\,Draw\ out_t = Inv\,Draw\,out_{t-1} + \text{Prod}\,Draw\,out_t - \text{Prod}\,GVZ_t \tag{14}$$

$$Inv\,FP_t = Inv\,FP_{t-1} + \text{Prod}\,mesh_t + Outsourcing_t - Stockout_{t-1} + Stockout_t - demand_t \tag{15}$$

```
model+=lpSum(inv_fp0)-lpSum(stockout0)+lpSum(allocation[9][0])+lpSum(allocation[2][0])-lpSum(demand[0])
-lpSum(allocation[0][0])+lpSum(allocation[1][0])==0

for i in range(1, Periods):
  print(lpSum(allocation[0][i-1])-lpSum(allocation[1][i-1])+lpSum(allocation[9][i])+lpSum(allocation[2][i])
  -lpSum(demand[i])-lpSum(allocation[0][i])+lpSum(allocation[1][i])==0)
  model+=lpSum(allocation[0][i-1])-lpSum(allocation[1][i-1])+lpSum(allocation[9][i])+lpSum(allocation[2][i])
  -lpSum(demand[i])-lpSum(allocation[0][i])+lpSum(allocation[1][i])==0
```

```
model+=lpSum(inv_GVZ0)+lpSum(allocation[8][0])-lpSum(allocation[9][0])-lpSum(allocation[11][0])==0

for i in range(1, Periods):
  print(lpSum(allocation[11][i-1])+lpSum(allocation[8][i])-lpSum(allocation[9][i])-lpSum(allocation[11][i])==0)
  model+=lpSum(allocation[11][i-1])+lpSum(allocation[8][i])-lpSum(allocation[9][i])-lpSum(allocation[11][i])==0

model+=lpSum(inv_draw_out0)+lpSum(allocation[7][0])-lpSum(allocation[8][0])-lpSum(allocation[10][0])==0

for i in range(1, Periods):
  print(lpSum(allocation[10][i-1])+lpSum(allocation[7][i])-lpSum(allocation[8][i])-lpSum(allocation[10][i])==0)
  model+=lpSum(allocation[10][i-1])+lpSum(allocation[7][i])-lpSum(allocation[8][i])-lpSum(allocation[10][i])==0
```

3.2.4.5 Inventory policy constraints for drawing, galvanizing, and transformation

To adhere to the defined inventory policies and prevent stock-outs or excess inventory, a loop is created where, for each period, it is established that the inventory of the respective process should be equal to the predefined amount of inventory on the initial list. A similar process is carried out for the "stock-out" variable.

$$Inv\,Draw\ out_t = Draw\ out\ inventory\ policy_t \tag{16}$$

$$Inv\,GVZ_t = GVZ\,inventory\;policy_t \tag{17}$$

$$Inv\,FP_t = FP\,out\;inventory\;policy_t \tag{18}$$

$$Stock\;out_t \leq Inventory\;Back\log_t \tag{19}$$

```
for i in range(0, Periods):
  print(lpSum(allocation[10][i])==lpSum(days_inv_draw_out[i]))
  model+=lpSum(allocation[10][i])==lpSum(days_inv_draw_out[i])
```

```
for i in range(0, Periods):
  print(lpSum(allocation[11][i])==lpSum(days_inv_GVZ[i]))
  model+=lpSum(allocation[11][i])==lpSum(days_inv_GVZ[i])
```

```
for i in range(0, Periods):
  print(lpSum(allocation[0][i])-lpSum(days_inv_fp[i])==0)
  model+=lpSum(allocation[0][i])-lpSum(days_inv_fp[i])==0
```

```
for i in range(0, Periods):
  print(lpSum(allocation[1][i])-lpSum(stock_out[i])<=0)
  model+=lpSum(allocation[1][i])-lpSum(stock_out[i])<=0
```

3.2.4.6 Non-negativity constraint

It is specified that the value of each of the variables in each period cannot be negative.

$$Hire_t, Fire_t, Workforce_t, Overtime_t, \mathrm{Prod}\,draw\;out_t, \mathrm{Prod}\,GVZ_t, \mathrm{Prod}\,mesh_t, \tag{20}$$

$$Inv\,draw\,out_t, Inv\,GVZ_t, Inv\,FP_t, Stock\,out_t, Outsourcing_t \geq 0 \tag{21}$$

```
for j in range(vd):
  for i in range(Periods):
    print(lpSum(allocation[j][i])>=0)
    model+=lpSum(allocation[j][i])>=0
```

3.2.5 Fifth step

Finally, the linear programming model is solved using the "solve()" function, which searches for the optimal solution to the problem, considering all previously defined constraints and objectives. The results are organized in a Dataframe that displays the model's solution, which means the values that each defined variable should take. Additionally, the total cost associated with the optimal solution is printed.

```
model.solve()
print("Solution Status=", LpStatus[model.status])
fila=[]
for Planning_Steel_wire in model.variables():
  fila.append(int(Planning_Steel_wire.varValue))
sol=pd.DataFrame(np.array(fila).reshape(vd, Periods))
sol.index=list(explica.values())
sol.columns=["Jul","Ago","Sep","Oct","Nov","Dec"]
print(sol)
print("Total Cost:", model.objective.value())
```

3.3 Results

After formulating the problem and subsequently optimizing it to minimize costs, the following aggregate plan has been established.

Total cost over the planning horizon: €1,880,655.

According to the obtained solution, there is no need for layoffs or additional hires. In other words, a constant workforce of eight employees will be maintained during the 6 months of planning, without the need for overtime, as shown in **Table 7**.

Similarly, with this workforce, there is no need to implement overtime or subcontracting. Throughout all the months, an inventory of both work-in-progress and finished products is maintained, which is why there are no shortages.

Additionally, as shown in **Table 8**, the following monthly production plan has been established for each of the processes involved.

As shown in **Table 9**, it is evident that the main cost component in the company's cost structure is related to material acquisition, characterized by a unit cost of €1 per

Month	Workforce	Workers to be hired	Workers to be fired	Overtime
July	8	0	0	0
August	8	0	0	0
September	8	0	0	0
October	8	0	0	0
November	8	0	0	0
December	8	0	0	0

Table 7.
Workforce and overtime required.

Month	Stock out (kg)	Outsourcing (kg)	Drawn inventory (kg)	Galvanized inventory (kg)	Finished product inventory (kg)
July	0	0	32,000	42,667	53,333
August	0	0	29,000	38,667	48,333
September	0	0	26,000	34,667	43,333
October	0	0	21,000	28,000	35,000
November	0	0	24,000	32,000	40,000
December	0	0	30,000	40,000	50,000

Table 8.
Stock-out, outsourcing policies, and inventories.

Month	Wire drawing production (Kg)	Galvanized production (Kg)	Mesh production (kg)
July	441,200	411,200	372,533
August	278,000	281,000	285,000
September	248,000	251,000	255,000
October	190,000	195,000	201,667
November	252,000	249,000	245,000
December	324,000	318,000	310,000

Table 9.
Production during the planning horizon.

kilogram. During the six-month planning period, it is anticipated that a total quantity of 1,733,200 kilograms of material will need to be acquired. This expenditure results in a total material cost reaching €1.8 million, representing 95% of the total planning cost.

Therefore, it is suggested that the company explore and make modifications to its supply strategies, considering possible alternatives such as evaluating new suppliers, initiating negotiation processes, or redesigning procurement policies and lead times for material.

This plan leads not only to the determination of optimal solutions but also to the achievement of more efficient resource management and effective planning. Furthermore, it provides organizations with the ability to make strategic decisions that enable them to thrive and reach their goals, effectively addressing the changing dynamics of the market.

4. Strategic sourcing

In today's business landscape, organizations face two significant challenges: increased volatility in demand variability and growing pressure to enhance their competitive advantage in the market. These sudden and unforeseen changes have given rise to conflicts throughout the supply chain.

High levels of shortages occur during periods of high demand, while excess inventory accumulates during periods of low demand. This lack of synchronization between supply and demand has led to increased costs, underutilized working capital, reduced responsiveness of companies, and dissatisfied customers.

In the face of this scenario, organizations must work toward finding a balance between supply and demand through effective coordination between the sales and operations/logistics departments. This is essential for increasing the company's profitability.

Firstly, on the sales side, strategies such as cross-selling, discounts, promotions, or more detailed pricing and revenue management can be explored.

Secondly, on the operations and logistics side, achieving supply smoothing can be accomplished through optimal inventory management, in terms of both sourcing and delivery frequency.

The fill rate, or service level, is a metric that relates to the attributes of reliability and responsiveness. Its main indicators focus on timely delivery, the right quantity, and appropriate quality [24].

Inventory availability plays a crucial role in demonstrating competitive service. This metric represents the probability of having sufficient stock to meet demand, thus minimizing the likelihood of running out of products during a specific period.

Availability is subject to constant evaluation, measured about a pre-established monthly target. However, this availability comes with a series of costs that can also impact competitive advantage. These costs include opportunity cost, which represents the investment in inventory instead of other activities; variable costs associated with storage; fixed costs related to storage; the cost of tied-up capital; costs associated with risk; and costs related to external services [25].

Therefore, a supply chain manager must ensure a balance between inventory availability and cost. This involves seeking the optimization of availability and frequency to maximize the profitability of the chain. This goal can be achieved through the implementation of procurement policies. Four tactics have been adapted from well-known inventory management models to respond to various demand behaviors.

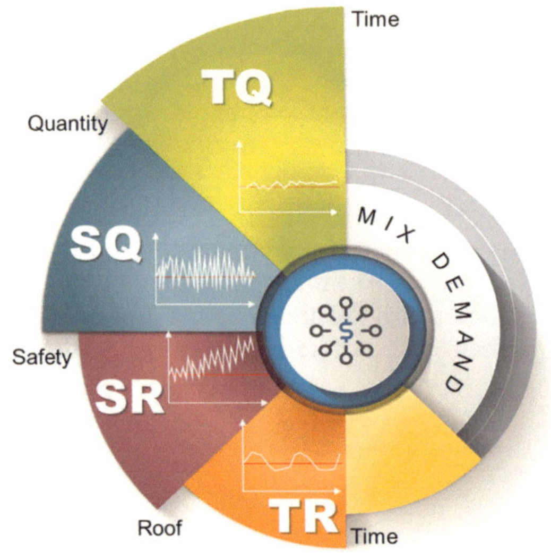

Figure 6.
Supply policies proposed.

Taking into account the variables T (lead time), Q (optimal order quantity), S (safety stock), and R (inventory roof), four procurement policies are adapted to address four different demand behaviors, as seen in **Figure 6** [26–29].

4.1 TQ policy

This model is based on the well-known concept of Economic Order Quantity (EOQ), which involves placing orders for a fixed quantity at a calculated frequency. The optimal order quantity, denoted as "Q", is requested at time intervals "T", which are calculated based on demand [26–29].

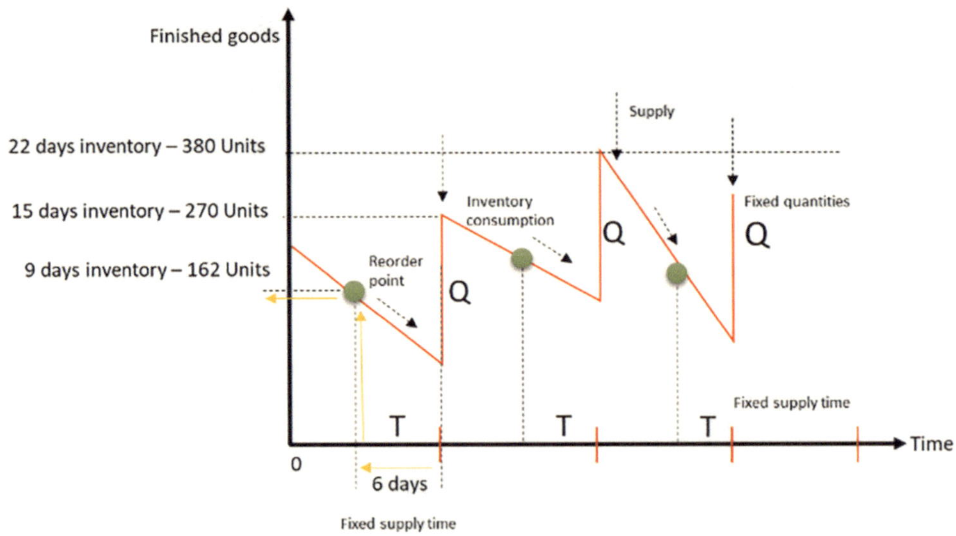

Figure 7.
TQ policy.

Abbreviations	formulas
T: Optimal supply time	T = Square root ((2*Cp)/(Cs*d)) T = Q′/d
Q': Optimal supply quantity	
Cp: Ordering cost	
Cs: Sustaining cost	
d: Average demand	

Table 10.
TQ formulas.

The demand exhibits a constant and predictable behavior, which enables the supplier to commit to a frequent delivery schedule. This strategy not only optimizes efficiency in transportation but also allows for more effective management of the quantity supplied.

Figure 7 provides a visual representation of this policy. The red line reflects the inventory behavior, where the sloping lines indicate consumption levels, and the vertical lines represent inventory replenishment points. By calculating the optimal time "T" for replenishment, it is possible to deduce the reorder point, which is indicated by the yellow markers.

With this information in mind, inventory consumption begins from time zero. As consumption reaches the green point, or the reorder point, an order is placed for a quantity "Q". This quantity is selected as the optimal balance between order costs and storage costs and will arrive in a period of "T", in this case, 6 days. Then, consumption occurs again, and the process repeats continuously as it touches the new reorder point. To perform the necessary calculations, the formulas are shown in **Table 10**.

4.2 SQ policy

This model aims to prevent inventory depletion situations by calculating a safety stock that should cushion the conditions of uncertainty that arise when demand shows variability and volatility [26–29].

Due to this nature, predicting and optimizing it is a challenging task. While it is possible to calculate the optimal quantity "Q", this model adapts to the supplier's lead times due to its condition.

Both demand and lead times are subject to variability, so the safety stock "S" will be determined by the variability of demand, lead times, and the level of service required by customers.

As illustrated in **Figure 8**, the inventory consumption process begins at time zero. When this consumption reaches the safety stock level, the order is triggered, as indicated in point 1. According to the agreement on the lead time with the supplier, the optimal quantity "Q" of the order arrives, as indicated in point 2, and inventory consumption continues. It is important to note that, due to demand variability, the slopes of the red lines representing consumption may be more or less steep.

Although this model is reactive and provides a quick response, sudden and aggressive demands can be challenging to cushion. As seen in point 3, this can lead to an inventory depletion situation, that is, a "stock-out", which is quickly resolved thanks

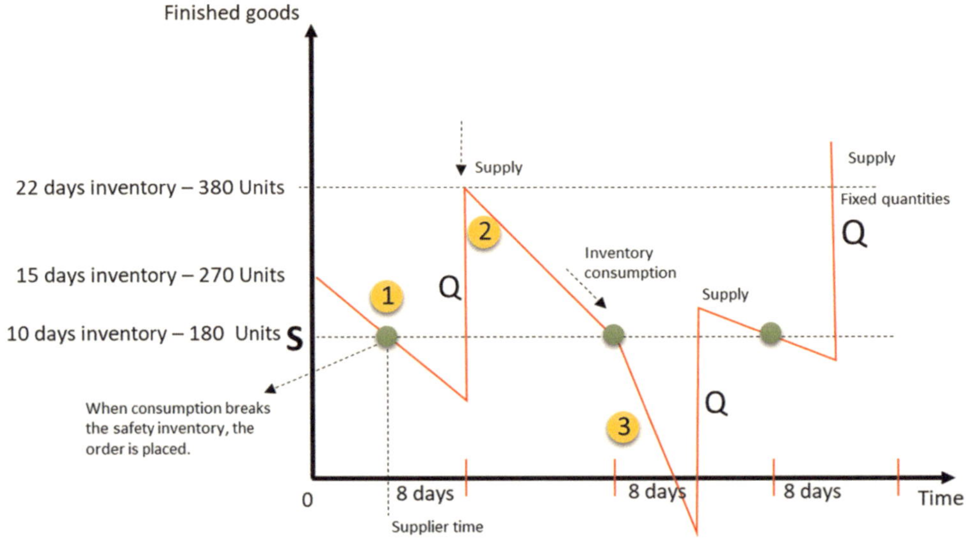

Figure 8.
SQ policy.

Abbreviations	formulas
Q′: Optimal supply quantity	Q′= (Square root ((2*Cp)/(Cs*d)))*d
Cp: Ordering cost	Sq = k * σ * (Square root (t))
Cs: Sustaining cost	k= NORM.S.INV[1] (P)
d: Average demand	P = Probability that they will not show up out of stock (Stock-out)
Sq: Safety stock for q	σ = STDEV[4] (Demand)
k: Safety factor	
σ: Standard deviation of demand	
t: Current supply time	

[1]*Excel functions.*

Table 11.
SQ formulas.

to the prompt response of the supplier, as the order was placed in advance. To calculate the data, it is necessary to review the formulas in **Table 11**.

4.3 SR policy

This combination aims to balance the inventory behavior, in both decrease and increase. In addition to demand volatility, it exhibits a trend toward growth or decrease. Although the lead time is not ideal, its high variability makes it difficult to calculate an optimal "Q". Instead, the order calculation is based on a maximum level "R" and the inventory level at period "n". [26–29].

Concerning **Figure 9**, inventory consumption starts in period zero with a total of 270 units, sufficient to cover a 15-day demand. When inventory consumption reaches the safety stock level, an order "Q_1" is generated, which is equal to the difference between the maximum level "R" and the inventory level at period "n_1", as indicated in point 1. The lead time is governed by the negotiated agreements with the supplier and is taken into account when calculating the safety stock.

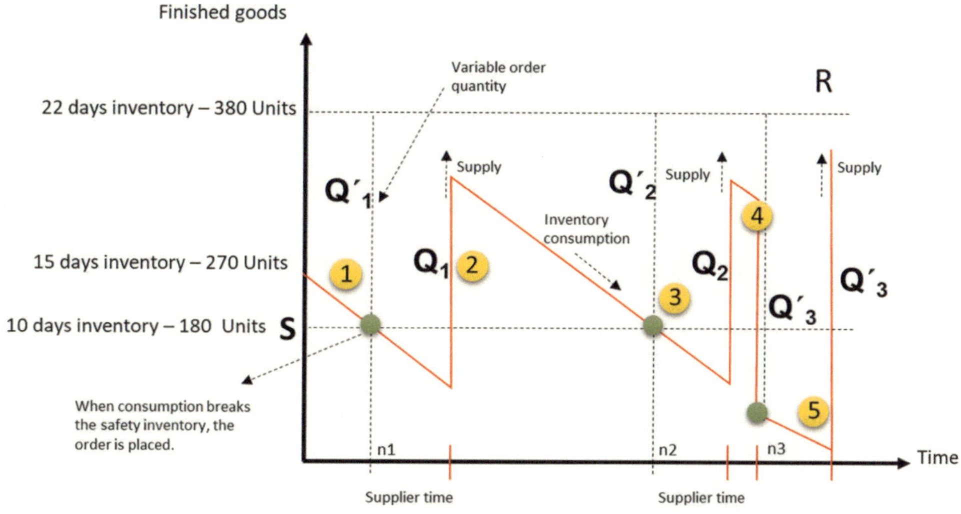

Figure 9.
SR policy.

At point 2, the requested quantity "Q'$_1$" is received, and consumption continues until the safety stock level is reached again, as shown in point 3. At this moment, a new order "Q'$_2$" is generated, and as agreed with the supplier, the material is delivered.

The inventory decreases again due to consumption, but at a specific point, a more significant consumption is observed in a single period. This can be interpreted as an unanticipated or special request, as shown in point 4.

This consumption causes the inventory to drop below the safety level, leading to placing an order for the quantity "Q'$_3$", which is the difference between the maximum level and the inventory at period "n_3".

After this significant consumption, the normal demand continues to deplete the remaining inventory, anticipating a stock-out. However, the supplier responds promptly with their delivery, which is reflected in the increase in inventory, as seen in point 5.

Since it is not possible to calculate an optimal order quantity "Q" or an optimal lead time "T" in this model, a quantity "Q'$_n$" must be calculated for each replenishment period "n". As a result, it is necessary to create a plan based on the maximum level "R", taking into account the formulas in **Table 12**.

Abbreviations	Formulas
SR: Safety stock for R	SR = d*t + k * σ * (Square root (t))
d: Average demand	d = Average D
k: Safety factor	k = NORM.S.INV[1] (P)
σ: Standard deviation of demand	P = Probability that they will not show up out of stock (Stock-out)
t: Current provisioning time	σ = STDEV[5] (Demand)
R: Roof	R_{SR} = d * t + SR
Q: Supply quantity	$Q'_n = R - i_n$
i: Inventory	$i_n = i_{n-1} - d_n + Q_n$
n: Period	

[1]*Excel functions.*

Table 12.
SR formulas.

n	1	2	3	4	5
d	d_1	d_2	d_3	d_4	d_5
R_{SR}	R	R	R	R	R
Q′ (Place order)	$Q'_1 = R - i_1$	$Q'_2 = R - i_2$	$Q'_3 = R - i_3$	$Q'_4 = R - i_4$	$Q'_5 =R - i_5$
Q (Order arrival)	$Q_1 = Q'_0$	$Q_2 = Q'_1$	$Q_3 = Q'_2$	$Q_4 = Q'_3$	$Q_5 = Q'_4$
i_{n-1}	i_0	i_1	i_2	i_3	i_4
i_n	$i_1 = i_0 + Q_1 - d_1$	$I_2 = i_1 + Q_2 - d_2$	$I_3 = i_1 + Q_3 - d_3$	$I_4 = i_2 + Q_4 - d_4$	$I_5 = i_3 + Q_5 - d_5$

Table 13.
Quantity SR planning.

As shown in **Table 13**, during the first period, there is a demand "d_1" along with an organization-set ceiling "R" and an initial inventory level "i_0". To determine the order quantity "Q'_n" for this period, the process begins by calculating the current inventory "i_1", which is obtained by adding the initial inventory to the received supply quantity "Q_1" in that period and then subtracting the demand for that same period "d_1". Next, the order quantity "Q'_n" is calculated, which is obtained by subtracting the upper limit "R" from the previously calculated initial inventory "i_1". This scheduled quantity will arrive as supply "Q_2" in the next period, following the agreed-upon lead times with the supplier.

4.4 TR policy

This model is designed to address seasonal demands by gradually adapting to smooth market fluctuations in a cyclical manner. It effectively cushions these changes using expected optimal lead times, thus avoiding the accumulation of unnecessary inventory. In this context, responsiveness is not as critical, although we should consider that there are peak demand moments that can impact product availability. The optimization of this model focuses on lead time and is accompanied by an upper limit to prevent inventory excesses [26–29].

As represented in **Figure 10**, inventory consumption begins at time zero. When consumption reaches half of the inventory, the order "Q'_1" is generated, as indicated in point 1. When the optimal time "T" is reached, the quantity "Q_1" from the order is received, as indicated in point 2, and the inventory consumption cycle continues. Because demand is seasonal, variations in demand rates are not as abrupt.

In this model, it is not feasible to calculate an optimal order quantity "Q", which leads to calculating a quantity "Q" for each period "T", but it is possible to achieve an optimum in the lead time "T". Consequently, it is necessary to develop a plan based on the maximum level "R", taking into account the formulas in **Table 14**.

As seen in **Table 11**, in the first period, there is a demand for "T_1", a calculated ceiling for "TR" (R_{TR}), and an initial inventory level of "i_0". To determine the order quantity "Q'_{T1}" for this period "T", the process begins by calculating the current inventory "i_1". This is obtained by adding the initial inventory to the received supply quantity "Q_1" in the same period and then subtracting the demand for that period "d_1".

Next, the order quantity "Q'_{T1}" is calculated. This quantity is obtained by subtracting the upper limit "R_{TR}" from the previously calculated initial inventory "i_1". The planned quantity "Q'_{T1}" will be delivered as supply "Q_2" in the next period (**Table 15**).

"T", which is the optimal time for replenishment.

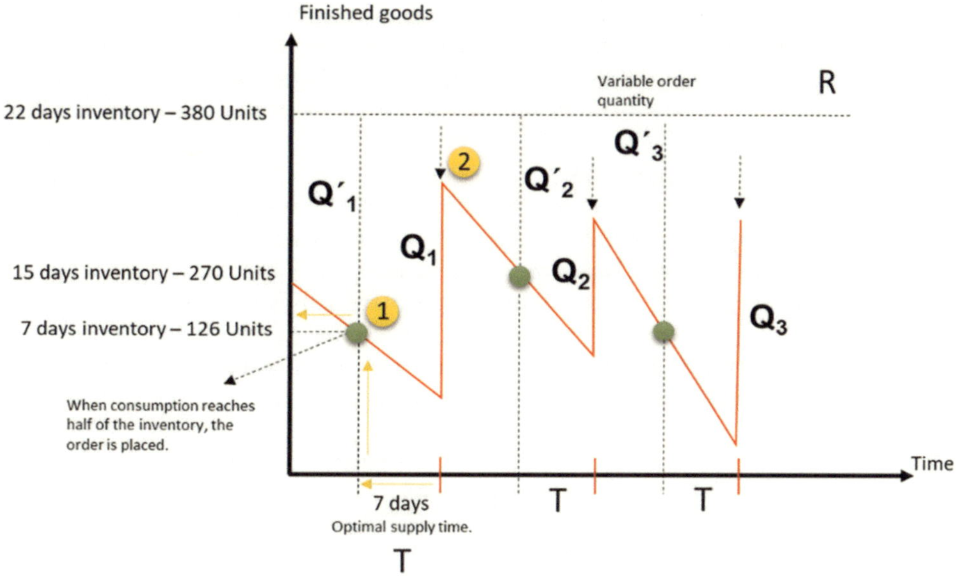

Figure 10.
TR policy.

Abbreviations	Formulas
T: Optimal supply time Cp: Ordering cost Cs: Sustaining cost d_{TR}: Average plus deviation demand for TR R: Roof Q: Supply quantity i: Inventory t: Current supply time	T = Square root ((2*Cp) / (Cs*d)) d_{TR} = Average D + Desvest D $R_{TR} = (d + \sigma) * t$ $Q'_T = R - i_T$ $i_T = i_{T-1} + Q_T - d_{TR}$

Table 14.
TR formulas.

T	1	2	3	4	5
d_{TR}	d_1	d_2	d_3	d_4	d_5
R_{TR}	R	R	R	R	R
Q´ (Place order)	$Q'_1 = R - i_1$	$Q'_2 = R - i_2$	$Q'_3 = R - i_3$	$Q'_4 = R - i_4$	$Q'_5 = R - i_5$
Q (Order arrival)	$Q_1 = Q'_0$	$Q_2 = Q'_1$	$Q_3 = Q'_2$	$Q_4 = Q'_3$	$Q_5 = Q'_4$
i_{T-1}	i_0	i_1	i_2	i_3	i_4
i_T	$i_1 = i_0 + Q_1 - d_1$	$I_2 = i_1 + Q_2 - d_2$	$I_3 = i_1 + Q_3 - d_3$	$I_4 = i_2 + Q_4 - d_4$	$I_5 = i_3 + Q_5 - d_5$

Table 15.
Quantity TR planning.

The implementation of these four supply policies originates from the need to provide a strategic focus to the supply and inventory operation, as suggested by Michael Porter regarding the bargaining power of suppliers. In many instances, suppliers gain a greater share of the value and, consequently, pass on the costs to other players in the chain. Powerful suppliers can exert pressure on an industry, especially if it cannot effectively control costs related to raw materials, particularly in terms of supply logistics. This largely depends on the concentration of suppliers in a particular sector [30].

The degree of supplier bargaining power determines how skillful the industry must be to avoid excessive dependence. In situations where the supply source is highly competitive, meaning that several suppliers are offering attractive prices, the industry has greater bargaining power. However, in cases where the source is oligopolistic, the industry loses its bargaining power and becomes increasingly dependent due to the scarcity of suppliers. Lastly, when the source is monopolistic, the industry becomes entirely dependent on that single supplier. In such a scenario, a supplier serving multiple industries will not hesitate to maximize its profits at the expense of the companies [31].

In situations where the number of industries requiring the services of a supplier is limited or when one of these industries represents a significant portion of the purchase volume, the supplier will seek to protect it by offering competitive prices, and discounts, and collaborating in training and promotional activities.

Summarily, it can be concluded that suppliers can exert a significant influence on the market, especially when the cost of materials and inputs represents a substantial percentage of the cost of the final product. Therefore, companies must be strategic and insightful in their planning, procurement, and storage processes.

5. Conclusions

Inventory planning will continue to evolve into new models and strategies, largely influenced by a country's conditions and its structural capacity to coordinate logistical activities, as well as the emerging way in which markets evolve. However, it is essential to keep in mind that inventory remains a significant asset and, therefore, a crucial part of an organization's working capital. As such, it must be managed strategically, especially in the procurement process.

Given the current evolution of the industry into its fifth generation, the information era becomes even more crucial in its management. In the realm of procurement, this aspect takes on particular relevance. Inventory planning supported by artificial intelligence has become an imperative necessity, as it establishes an essential link between demand behavior and efficient industry responsiveness.

Inventory management becomes an essential strategic resource for organizations, as it plays a critical role in optimizing operations and a company's profitability, as well as customer satisfaction. A thorough understanding of the influence of inventory in all areas of a company, from operations to finance, enables organizations to appreciate the value it represents as an asset. This, in turn, empowers them to develop various strategies aimed at generating benefits, maximizing operational efficiency, and achieving greater profitability.

It is essential to recognize that while inventories are an asset for companies, they also represent a capital investment. Inadequate inventory management can trigger financial problems or reduce service quality. Therefore, it is crucial to achieve a balance between inventory levels, their associated costs, and productive capacity.

Currently, the market is characterized by constant dynamism and uncertainty, representing significant challenges for organizations in their planning strategies.

DOI: http://dx.doi.org/10.5772/intechopen.114025

Therefore, aggregate planning emerges as a fundamental tool that allows for strategic decision-making, considering both demand volatility and the organization's production capacity. However, companies must understand that to achieve effective results when implementing this model, it is necessary to foster collaboration throughout the supply chain.

An aggregate plan, the integration of all supply chain actors, and effective inventory management not only ensure flexibility and adaptability to market changes but also enable agile and flexible operation. This becomes even more crucial in highly competitive and volatile markets. Strategic strengthening in operations management is no longer a choice; it becomes imperative if an organization aspires to enhance its competitive advantage.

For any professional working in logistics or supply chain areas, mastering mathematical models and the ability to program algorithms become essential. Success in decision-making lies in the quality of information and, consequently, in the precise modeling of logistical or planning problems. This allows for accurately predicting operational behavior at a competitive cost and anticipating demand, ensuring adequate service levels at a competitive cost.

Inventory management is a fundamental part of an organization's operations and has a significant impact on its profitability and efficiency. As the industry evolves into fifth-generation models and the information era becomes more crucial, inventory planning supported by artificial intelligence becomes a necessity to stay competitive and adaptable to market changes.

Author details

Andrés Felipe Santos Hernández and Natalia Camacho Franco*
Escuela Colombiana de Ingeniería Julio Garavito, Bogotá, Colombia

*Address all correspondence to: natalia.camacho@mail.escuelaing.edu.co

References

[1] World Economic Forum. The Global Risks Report 2023. Cologny/Geneva: WEF; 2023

[2] World Economic Forum. Chief Economists Outlook. Cologny/Geneva: World Economic Forum; 2023

[3] World Bank. Trade Logistics in the Global Economy. Washington: WB; 2023

[4] Departamento Nacional de Planeación. Encuesta Nacional Logística 2020 [En línea]. Colombia: Departamento Nacional de Planeación; 2020. Available from: https://plc.mintransporte.gov.co/Portals/0/News/Encuesta%20Nacional%20Logi%CC%81stica%202020.pdf?ver=2021-09-24-211753-007 [Último acceso: Abril 25, 2022]

[5] Consejo privado de competitividad. Informe Nacional de Competitividad. Bogotá: Puntoaparte; 2023

[6] Portafolio, Portafolio. 2023. [En línea]. Available from: https://www.portafolio.co/negocios/empresas/el-sector-retail-se-esta-moviendo-entre-la-inflacion-y-la-necesidad-de-mas-tecnologia-578906 [Último acceso: August 5, 2023]

[7] La Republica, La Republica, Comercio. 2023. [En línea]. Available from: https://www.larepublica.co/empresas/fenalco-analizara-el-futuro-del-retail-junto-a-deloitte-para-plantear-soluciones-a-crisis-3552567

[8] Deloitte group. Perspectivas de la industria de. Londres: Deloitte group; 2023

[9] Portafolio, Portafolio. 2023. [En línea]. Available from: https://www.portafolio.co/negocios/empresas/industria-logistica-ejes-de-trabajo-para-el-sector-durante-2023-576901

[10] World Bank, World Bank. 2023. [En línea]. Available from: https://www.bancomundial.org/es/news/press-release/2023/01/10/global-economic-prospects

[11] Coyle JJ, John Langley C Jr, Novack RA, Gibson BJ. Administración de la cadena de suministro. Mexico, D.F: Cengage Learning Editores; 2013. p. 674

[12] Andersson MM. Exact analysis of one-warehouse-multiple-retailer inventory systems with quantity restricted deliveries. European Journal of Operational Research. 2023;**2023**(309):1161-1172

[13] Martínez Magda Cejas, Bayas Irma Yolanda Garrido. Inventory management as a strategic. Scientific e-Journal of Management Science. 2017;**37**:109-129

[14] Olli-Pekka H. Role of inventory and assets in shareholder value creation. Expert Systems with Applications. 2020;**X**:1-7

[15] Koumanakos DP. The effect of inventory management on firm performance. International Journal of Productivity and Performance Management. 2008:7(2):383-392

[16] Kontuš E. Management of inventory in a company. Ekonomski vjesnik. 2014;**27**(2):245-256

[17] Brigham EF, Ehrhardt MC. Financial Management Theory & Practice. Florida, United States: South Western Cengage Learning; 2020

[18] Edwin Sitienei FM. The effect of inventory management on profitability of cement manufacturing companies in Kenya: A case study of listed cement manufacturing companies in Kenya, International Journal of Management and Commerce Innovations. 2016; **3**(2):111-119

[19] Chopra S. Administración de la cadena de suministro Estrategia, Planeación y Operación. Mèxico: Pearson education; 2013

[20] Smith Quintero R, Espinal A. Correa, Aristizábal JA. Un enfoque de analisis multiobjetivo para la planeación agregada de producción. DYNA. 2004; **71**(141):15-27

[21] Boca TPM. The Wire Drawing Process Simulation and the Optimization of Geometry Dies. Cluj-Napoca, Romania: Elsevier; 2017. pp. 187-192

[22] Berto RAM. Mechanical behavior of hot-dip galvanized welded steel under. Procedia Structural Integrity. 2017;**3**: 135-143

[23] IDU. Guía práctica de galvanizado en caliente. Bogotá: ANDI; 2020

[24] APICS. Supply Chain Operations Reference Model 12 Version. Chicago: APICS; 2020

[25] Gao Dandan, Wang Nengmin, He Zhengwen, Zhou Li. Analysis of bullwhip effect and inventory cost in the online closed-loop supply chain. International Journal of Production Research. 2023;**61**(20):6807-6828

[26] Rossini Matteo, Kundu Kaustav. Case study on simulation analysis. Management Research Review Communication of Emergent International Management Research - Emerald. 2022;**46**(4):508-533

[27] Krajewski L, Ritzman L. Administración de operaciones. México: Pearson; 2013. pp. 462-489

[28] Santos CJ. Digital transformation of planning in the pharmaceutical sector. In: Changing Tides: The New Role of Resilience and Sustainability in Logistics and Supply Chain Management. Berlin, Germany: ECONSTOR; 2022. pp. 267-293

[29] Chopra S. Administración de la cadena de suministro Estrategia, Planeación y Operación. Mexico: Pearson; 2015. pp. 314-347

[30] Porter ME. Five forces that shape strategy. Harvard Business Review. 2008;**Enero**:82-83

[31] Coyle JJ, John Langley C Jr, Novack RA, Gibson BJ. Contratación de materiales y servicios, de Administración de la cadena de suministro. Pensylvania: Cengage; 2015

Chapter 5

MGSC Project Process Management

Vladimir Križaić

Abstract

Many scientists have worked on the mathematical definition of the S-curve, including mathematicians, statisticians, economists and engineers. The cash flow or S-curve defines costs and revenues in a planned way using a cumulative curve that simulates the existence of the project. There is always a deviation in various process activities and the consequences are financial, which puts the project stakeholders in a difficult position as it usually involves large investments. Therefore, the case of financial flow forecasting is determined by the modification of the Gaussian S-curve (MGSC) through scientific project management, which is very useful for all participants in the production process. The MGSC solves the problem of defining the initial state of the project in terms of the duration and cost interval of the project. However, there is the problem that the duration end and the project costs increase or decrease differently during the realisation of the project, which is defined by different process activities. Simplified assumptions within the process activities and the risk method complement the differentiated development of the process and create a dynamic simulation of the project, that is daily project management. This enables timely regulation of the execution system, that is the management of process projects.

Keywords: cash flow, process projects, daily monitoring, MGSC, organisational differential

1. Introduction

The rapid development of project management (PM) has been made possible by the rapid development of business process management (BMP) technology. New CAD and SMART technologies and methods within integrated information systems (IIS) contribute to the daily monitoring [1] of business and project management systems. The key processes in these new methods are the planning and regulation of the project system, that is the planning process group and the monitoring and control process group. Standardisation technology, however, does not follow the third and fourth industrial revolutions; hence Professor Rex's phrase, "If you know technology, you know how to programme", has been proven true in practice. It was programming that enabled the development of the method of dynamic structural programming (DSP), which, together with operational research as a tool for optimal project management, transforms the static project of building organisation (POB) [2] into a dynamic POB, that is dynamic project management processes (PPM). The implementation of an appropriate enterprise resource planning (ERP) system [3] is a key investment that usually ties up enormous resources and can have a significant impact on the future competitiveness and operation of the company. In today's practice, the last planning system (LPS), followed by location based management (LBM) and 4D modelling in space with a time component, then planning on work cycles [4] (PCW) and the traditional approach based on critical vay metoda (CPM) in Microsoft Project (MSP), Primavera or SuperProject. Any project realisation is mainly production. It is a function of technology, organisation, management and information and communication technology (ICT). The technology uses differential equations, while the other components run behind the mathematical modelling of these processes. Therefore, software models and simulations, such as BIM and Digital Twins, appear today as tools that aim to complement the missing decision-making processes. This results in the need for vector [5] or parametric [6] modelling and the transformation of static POB and project management documents (PDM) into dynamic scientific documents. The sustainability of BIM requires the modelling of n-dimensional dimensions of nD buildings as components of dynamic structural programming (DSP) or application programming interface (API). Thus, for the process engineer, the focus is on the sustainability of the traditional criteria of structural integrity, constructability and cost. However, today's ITC digital technology also provides for the automation of the work for which the GPS technology was developed, which converts manual records into automatically collected data in the field and sends it automatically to the central data processing in the company's IIS system. However, the key to solving the problem lies not only in automating existing administration, but also in standardising production standards (construction, mechanical and electrical) and cost items. Thus was born the DSP [7] dual COD method, which generates the iterative differential of cyclical production processes. Here, mathematical and cybernetic methods are used to transform organisational structures into differential organisational structures as the main components of the artificial intelligent (AI) system. Thus, API and DSP applications are useful links to launch add-ons and external programmes to extend and optimise BIM. Moreover, with robotisation, technology has risen in the organisation, so it is necessary to build dynamic standard models with vector planning [8] and improve them by optimising production processes [9]. Thus, model standardisation of business processes eliminates problems and seeks to optimise the economic and organisational elements in the business system defined by the financial cash flow curve. This process of bringing organisational solutions closer to technological achievements in order to eliminate their gap leads to the creation of the financial MGSC as a simulation element of the production model of modern cyber-physical systems (CPS) [10]. The goal of MGSC automation is softwareisation, that is model standardisation of production processes using mathematical and statistical methods to replace routine and hazardous operations with robots and reduce financial risks to an optimum for investment. The contribution of the management of the MGSC method lies in the adaptation, refinement and solution of problems in processes and methods in project management. Thus, in the first place, the transformation of a static process into a dynamic financial process with a curve linked to resource costs leads to a vector simulation MGSC that allows a technological record of the work or process that largely guarantees the quality of the work, that is integrated information quality management [11]. The objectives of managing the MGSC method are to promote a different way of recording the process as a basis for standardising mobile management to achieve management standards according to the methodology of project management (MP) [12]. Then the creation of a new vector S-curve as the basis for mobile management of management project process groups, especially in the group of planning the implementation and monitoring and controlling or regulating the cybersecurity system [13]. This, in turn, leads to a competitive price and the desired product quality, with the possibility of simulating management. A team manager who has daily dynamic data in smart technology becomes a manager of daily monitoring [14] of the total budget (EAC) and estimating the variance of the total duration of the project (BAC).

2. Vectoral organisation and technology on production process

The postmodern organisation begins with the introduction of information technology and defines the organisation systematically and procedurally with input resources (R) and output system products (S) (**Figure 1**).

Such a process organisation creates business resource planning (ERP) through the Information Business System of the American company SAP and the Japanese JIT and TQM systems.

The systemic approach is defined by the general theory of the system and strives for all resources of the system to be a function of the dynamics of the environment and feedback or control. All this is based on information moving from the classical vertical hierarchical flow to the horizontal process flow of communication. Thus, the theory of dependency on input resources for the system emerges. Then, business reengineering management (BMR) moves towards thought management of enterprises by creating a creative team leader instead of a command and control leader. MP [16] is done through three, four and five management functions with continuous decision-making that is constantly loaded with risk, as an uncertain event or condition that may occur in the future and disrupt a project goal, temporally, financially or qualitatively, reducing it to a tolerable risk. For the implementation of the project, it is important to prepare the work with the creation of a dynamic POB (**Figure 2**) [17] or PDM.

It can be seen that static standards and calculations dominate production management and the bid design is very inaccurate both financially and qualitatively as a product. However, with dynamic, vector or parametric modelling of the standard and calculation, this problem is greatly reduced. Therefore, vectorial and parametric models of standards are created, which are used to create digital twin systems for graphical and analytical representation of production.

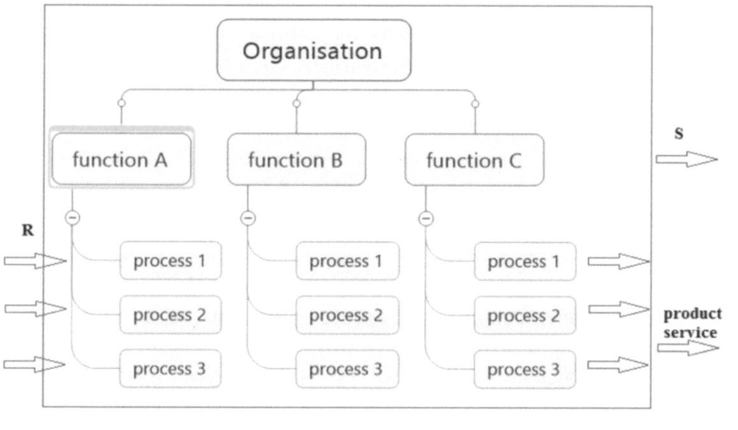

Figure 1.
Systemic-processual [15].

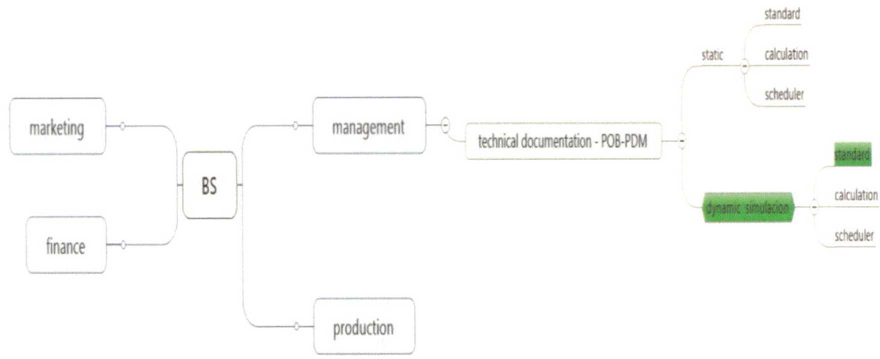

Figure 2.
Static and dynamic POB.

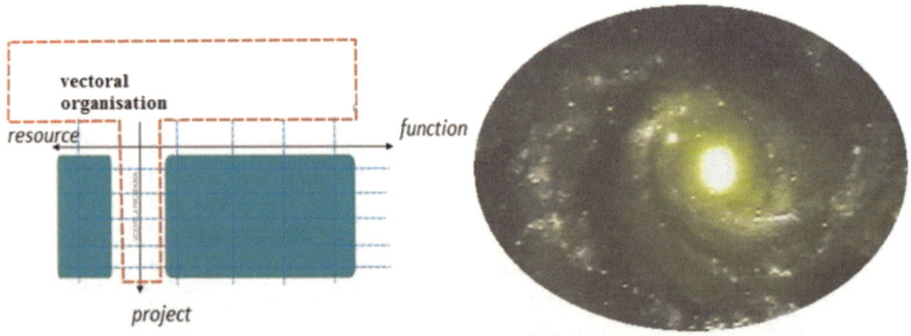

Figure 3.
IIS construction company with cyclo-space vector structure.

2.1 Vectoral organisation on production

Every business system consists of 4 functions, and within the management or administrative function, the most important sub-function is the planning of the production process and the regulation of the production processes, which are in the function of technology, organisation, management and ITC system with the daily management of the production or business system. Nevertheless, planning in construction is at a low level due to outdated standards. Therefore, the introduction of dynamic or parametric vector standards is recommended to increase the accuracy of planning and thus the cost of projects. As a result of relational databases, a vector organisational structure [18] of the company is created, which is introduced into the matrix organisational structure with a third axis, namely resource variables, which is stored in the database with histograms and allows access to data up to the resource level (**Figure 3**).

By elaborating the vector structure of the business system down to the level of the construction tender process, through the tender and site and situation documentation, to the level of operations (1 collection of tenders, 2 technological parameters, 3 planning parameters, 4 costing and others), that is working with the same human resources, a process elaboration of the management of **Figure 4** is created in five levels, that is processes. Thus, the POB becomes a PDM that goes through five levels, that is a group of processes that are roughly systematised in both systems.

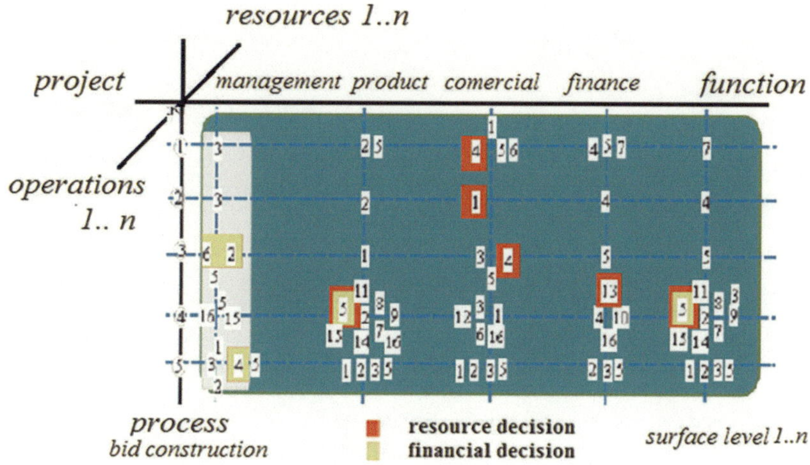

Figure 4.
Vectorial project-process-operation-resource-oriented structure reduced to the level of the function-process-surface.

2.2 Vectoral norm on technology production

In the same way, the 4D model of the norm, or the vector norm, was developed from the static norm. The materials and basic resources in **Table 1** are a function of the constructive and geometrical characteristics of the building project using the described technology.

Material consumption is defined by connecting the designer's equations with the effects of resources reduced to the dimensions of the design documentation of GK Međimurje formwork technology defined by the equations in the monogram with units of measurement in cm, m, kN. Labour costs By creating standards according to workers' occupations, the discrete data of labour standards are shortened and the model is obtained by defining the functional relationship of resource and time standards in relation to the dimension of construction or labour. By analysing and systematising the old standards, the three-dimensional variables of construction, technology and organisation can be used to define the vector standard hypothesis [19] Eq. (1).

$$VN = f(kd, ko, k, r) = kd_i ko_i f(k, r) \tag{1}$$

Normative time can be represented as a resource effect in a three-dimensional formula. The variables are elements of project construction (Xc) and technology (Y*r*) defined by the categorised variable of this organisation (Z*o*) (**Figure 5**).

Constructive elements	Geometric charact.	cm	cm^2	I	W	M max	fm	R
	b	h	F	cm^4	cm^3	kNm	kNm^2	kN
Bed	10	12		1440	240	2.4	10,000	7.2
DOKA H20	8	20		3850	385	5	13,000	11
Tubular scaffolding	5		498	12.4	5.14	0.108	21,000	10.33

Table 1.
Construction characteristics carpentry technology GK Međimurje 30 years ago.

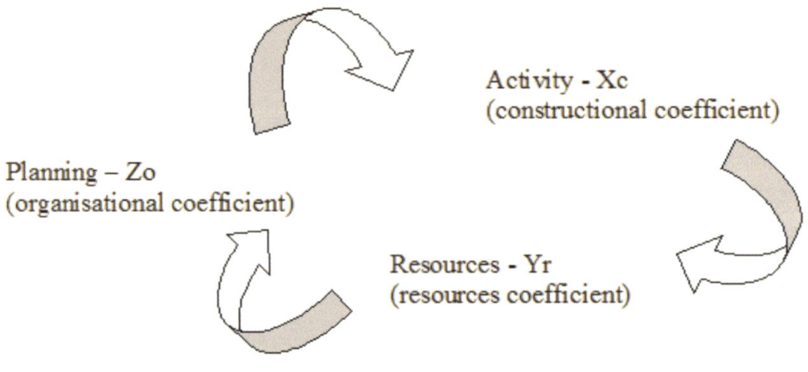

Figure 5.
Idea for the creation of a norm vector in the building industry.

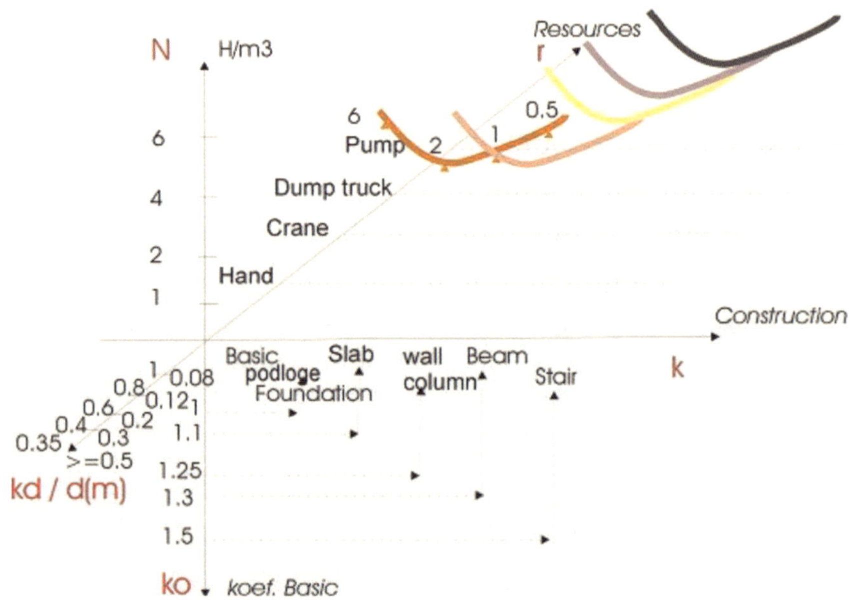

Figure 6.
Performance and vector norm for the concreting of buildings.

Thus, using different technologies as examples, tables were created on the function of the construction element of the execution (k) and the dimensions of this execution element (k*d*) and on the resource (r), that is the execution technology, and on the organisation with the definition of the basic type of the construction element (k*o*) as an organisational unit (**Figure 6**).

3. Cash flow of the project

The definition of the S-curve has been achieved by integrating management software, that is using ODBC databases or newer XML internet technology with the software MS Project. The system allows monitoring of resource efficiency by activity and on a daily basis through the presented data model (**Figure 7**).

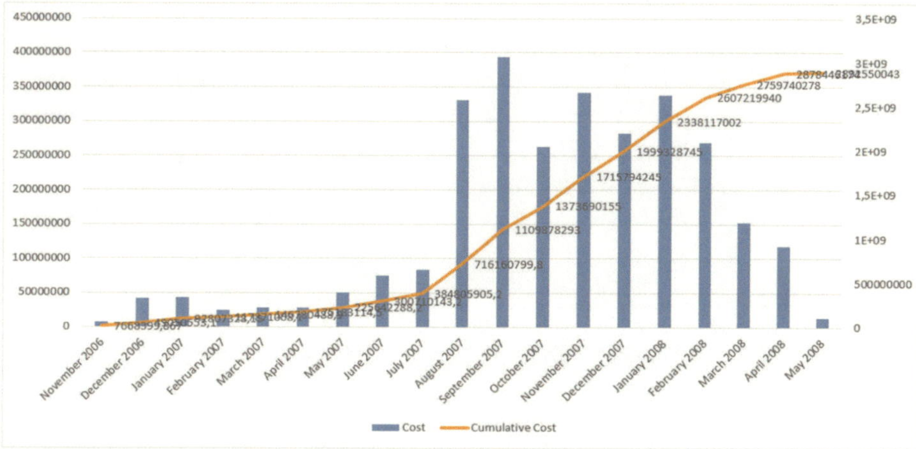

Name	Remaining Cost	Actual Cost	Cost	ACWP	BCWP	BCWS
1 1 ZAGREBAČKA STANOGRADNJA SESVETE	2.892.550.043,04 kn	0,00 kn	2.892.550.043,04 kn	0,00 kn	0,00 kn	0,00 kn

Figure 7.
Report on project monitoring and cash flow.

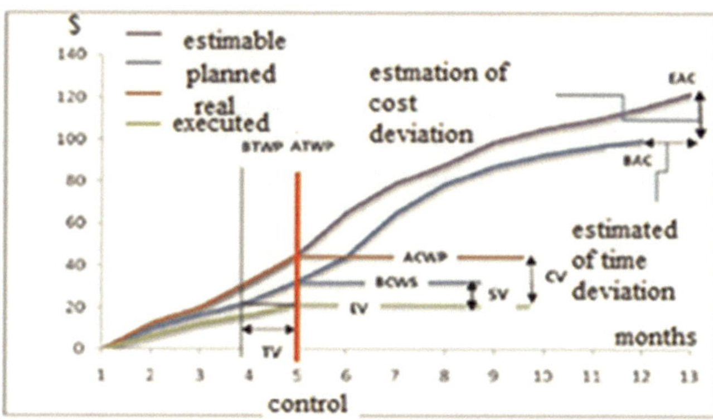

Figure 8.
Cash flow [20].

Alternating monitoring of normative and financial indicators has been made possible, which allows increasing, that is streamlining, the resource control of projects previously monitored on a periodic financial basis to define the estimate of the deviation of the EAC and the estimate of the deviation of the project BAC from the planned cash curve, for which we do not have a functional link but an assessment (**Figure 8**).

3.1 Planning the project process with MSP

An overview of the S-curves in MSP shows that distribution and cumulative curves are defined with attempts to define equations, but the nearest exponential curve does not satisfy the given S-curve (**Figure 9**).

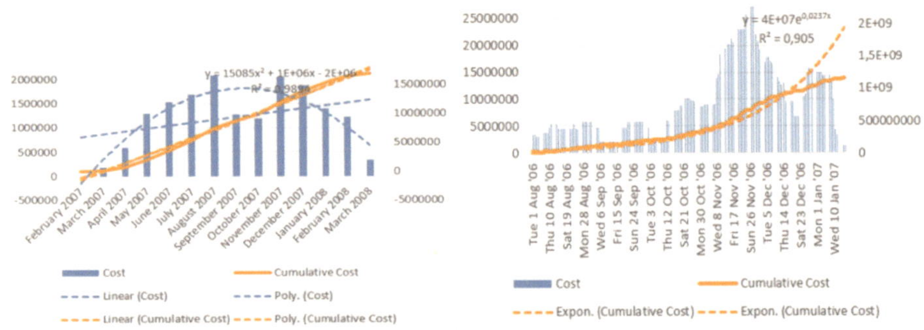

Name	Remaining Cost	Actual Cost
1 PROJEKT ELEKTRODA	16.722.619,20 kn	0,00 kn

Name	Remaining Cost	Actual Cost
1 PROJEKT HLADNJAČA,	1.152.775.353,61	0,00 kn

Figure 9.
Two specific projects using the S-curve report of the MSP software.

Both examples are different and show the monthly distribution of finances with a circular or approximately modified Erling density distribution similar to the future normative charts. Thus, MSP defines the cumulative quadratic and exponential curve of finances by a general linear, polynomial, quadratic or exponential equation Eq. (2).

$$y = ax^2 + \mathrm{b}x + \mathrm{c};\ \mathrm{y} = \mathrm{a}e^{bx} \tag{2}$$

Based on the detailed presentation of the reconstruction project of the Zagreb Maribor motorway bridge over the Sava in Zaprešić, the part over the railway and the Sava river, a static and dynamic presentation of the project was staged using the IIS system with the MS project presentation of the given project. It is obvious that the given IIS system has a problem with static normalisation. For example, some elements of the offer have no norm, so they are covered with an unanalysed resource and never defined exactly, and we want to define the resources in the project exactly. For a given project, we create a static ABC curve with the direct costs of the project, and we get a summary of the non-indirect costs with the profit that defines the rest of the finances of the project. Thus, the total investment of 26.4 million gives 17.4 million direct costs DT and other indirect costs IT with profit P Eq. (3).

$$UP = DT + IT + P \tag{3}$$

ABC Curve direct resource costs are defined in the order of the minimum (**Figure 10**, **Table 2**).

It is noticeable that the leading payback is pumping robots, asphalt, labourers, oil, machinists and concrete, which account for up to 80% of the price. Therefore, these resources should be influenced. To maintain a dynamic state of resources in the project, the labour and machinery resources are used to define the duration of the project. Thus, the labour is about 3 million and the machine is about 8 million kn, which gives the S-curve without material and other (**Figure 11**).

The S curve is incomplete because the transfer of resources has not been completed in full except for time, without which there is no Gantt chart. But for an MSP, moving the activities and transparency of the Gantt chart is a real routine. The direct costs of resources are visible in the histogram and are made up of labour R, material M and machines S Eq. (4).

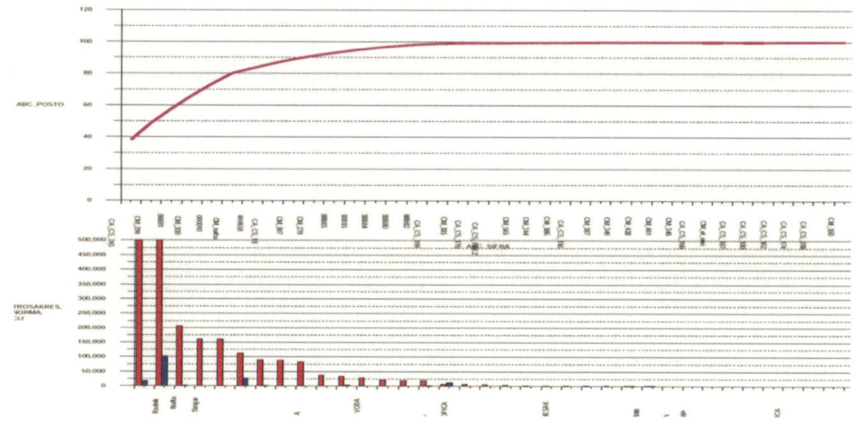

Figure 10.
ABC curve direct resource costs.

Bid construction resource	111,743-5R name	MU	Q	Max price	Costs	%	ABC	%
CA_CS_243	Robot	h	3291	2000	6,583,968	38	6,583,968	38
CM_299	Asphalt	T	4198	443	1,861,477	11	8,445,445	48
88,881	Indefinite	HRK	1,550,739	1	1,551,739	9	9,997,185	58
CM_300	Asphalt	T	4198	345	1,451,458	8	11,448,643	65
10	Worker	h	22,354	57	1,281,511	9	12,730,154	73
CM_nafta	Oil	l	166,892	7	1,168,247	7	13,898,402	80
650	Machinist	h	6623	80	529,907	3	14,425,310	81

Table 2.
ABC Curve direct resource costs.

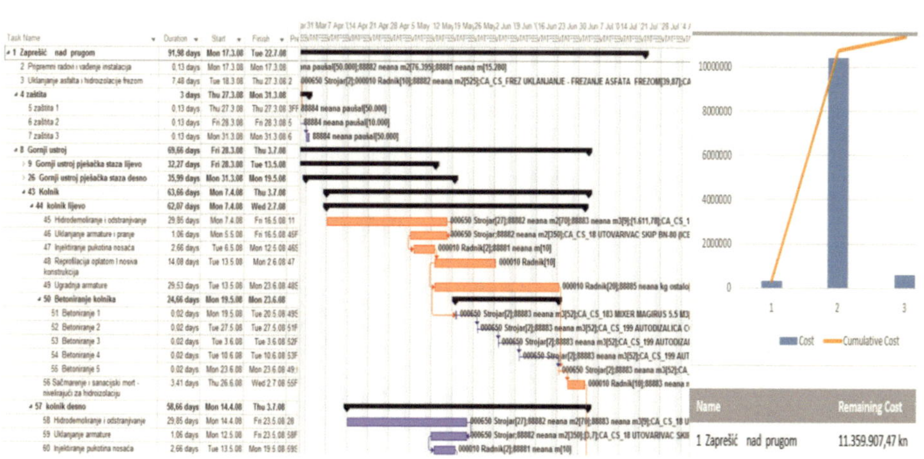

Figure 11.
Concrete project with the S-curve report of the MSP software of a bridge over the Sava and part of the railway line.

$$T = R + M + S \tag{4}$$

The equation Eq. (5) for the resource in the production time is a linear function of the product of the direction coefficient of the direction (a) or y' and the production time t.

$$R = M = S = at = y'\mathrm{t} \tag{5}$$

By reading from the Gantt chart or histogram, the resource R is located in the third dimension of the production budget; therefore, the equation takes the form of a sum Eq. (6).

$$R_j = \sum_{i=1}^{n} a_r t_{ri} \tag{6}$$

3.2 Planning the project process with MGSC carve

More recently, probabilistic forecasts of project performance and the use of stochastic S-curves with a stochastic S-curve generation software package and a simulation approach have defined the dispersion variance of finances and time of the MGSC project as a function of the density distribution of cost and time. Expected monetary value (EMV) or S-charts are widely used today and are complemented by functional MGSC [21, 22] Eq. (7), (**Figure 12**).

$$skvGr(x, T) = \lambda kv \cdot \int_0^x \frac{1}{(a \cdot T + b) \cdot \sqrt{2 \cdot \pi}} \cdot e^{\frac{-(x-\mu)^2}{kv \cdot (a \cdot T + b)}} dx \tag{7}$$

The levelling and fitting of the curves, that is the modification of the Gaussian curve, refers to the introduction of a constant kv parameter in the value of 10,000 units. The value of the investment in the reconstruction of the Zagreb-Zaprešić Bridge, which can be seen on the MSP fall stream, is HRK 2,640,000,000. In the software budget, the cost axis is in thousands, so T = 26,400 × 10^3. The number of months working on the project is n = 5, which defines the variable x, and the expectation of μ from the variable x is x/2. The constant variable kv is 10,000, and λkv is taken 100 times larger than the standard value and is 65,000, while the coefficient b of the standard value is 4 × 10^{-3} and has a slightly increased value of 3,81 × 10^{-8}.

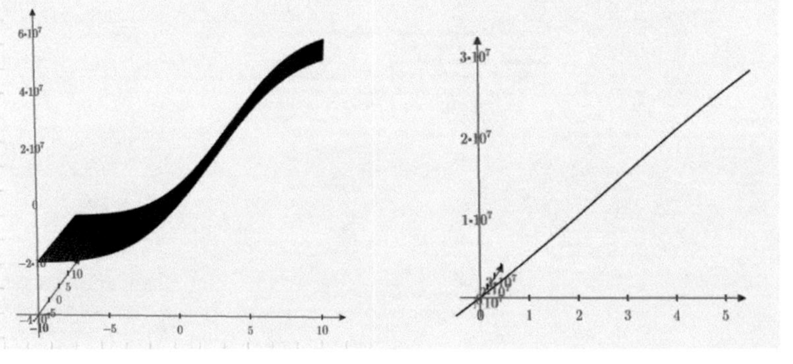

Figure 12.
MGSC project Zagreb-Zaprešić within –n to n and 5 months.

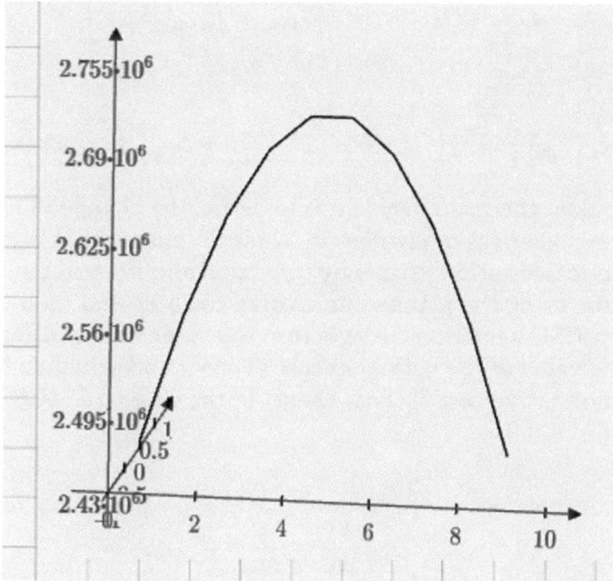

Figure 13.
Simulation of the density distribution of financial plan situations with MGSC for the Zagreb-Zaprešić project.

Considering that the S-curve has half monthly rates for the investment distributed within 2.4–2.7 million time-tenths, that is every half month, this is the financial differential, that is the density of the distribution of the financial plan situations in the project time (**Figure 13**).

4. Organisational differentiation in MGSC with the method DSP for production companies

The model standardisation of the cost description of items, that is the common harmonisation of the item descriptions in the offer, that is the standards, creates a unique code to define the construction product with the method DSP, that is the code DSP Eq. (8). The method DSP creates a structural organisational differential by levels with the possibility of elaboration at the movement level.

$$DSP\ COD: = [f_n([A|R|D||]) = f_{n+1}([A\ |R|D||])] \quad (8)$$

Goods as a function of the variables A-activities, R-resources and D-dimensions of design and resource performance. In this way, product datasets are formed by levels and distribution of complex processes to procedures, and construction production datasets are modelled by combinatorics and connected by standardisation of the model. By converting the code DSP into the structural code of the MGSC curve, an organisational or financial differential is created by a three-dimensional structural model defined in the i, j, k structure S_ijk Eq. (9), and by replacing the structure S with the MGSC curve, the MGSC DSP COD Eq. 10 based on iterative numerical methods [23] (Eqs. 9–11), which generates the future state of structure n + 1 using the existing state of structure n and the differential Δn + 1 shift of the function or structure.

$$DSP\ COD: = \left[f_n\left(\left[S_{ijk}\right]\right) = f_{n+1}\left(\left[S_{ijk}\right]\right)\right] \quad (9)$$

$$x_{n+1} = f(x_n) \quad (10)$$

$$MGSc\ DSP\ COD: = \left[f_{n+1}\left(\left[skvGr(x,T)_{ijk}\right]\right) = f_n\left(\left[skvGr(x,T)_{ijk}\right]\right)\right] + \Delta f_{n+1}\left(\left[skvGr(x,T)_{ijk}\right]\right) \quad (11)$$

In implementation, the main problem is to define the change in the EAC, BAC S-curve caused by various risks or changes in labour intensity, that is resource use. This is solved by numerical iteration, that is by induction and iteration of the cost estimating equations Tproc using the planned normative costs Tp and their derivation. The derivative of the MGSC valuation curve is thus the value of the difference between EAC, BAC and the value of the valuation cost (Tproc) and schedule (Tplan) functions, that is the normative cost T (Eqs. 12–14) of the standard MGSC curve plus the differential cost T.

$$T_{x+1} = T_x + \Delta T_{x+1} \quad (12)$$

$$T' = \frac{\Delta T}{\Delta x} = \frac{\Delta T}{\Delta t} = y' \quad (13)$$

$$T_{x+1} = skvGr(x,T) = \lambda kv \cdot \int_0^x \frac{1}{(a \cdot T + b) \cdot \sqrt{2 \cdot \pi}} \cdot e^{\frac{-(x-\mu)^2}{kv \cdot (a \cdot T + b)}} dx + T'\Delta x \quad (14)$$

At intervals of dozens of S-curves creating intervals of half a month within the 5 months of the project. This is how we define the increment ΔT. By numerically deriving the given curve skvGr(x,T) or Tp, the derivative of the function is created by tens parts of the function and approximated by Newton's tangent method [24], with the possibility of using steps of the degree of rotation or the intensity of resource consumption. The derivative function MGSC of the planned curve is defined with the Gaussian method of the sum of least squares.

$$x = \begin{bmatrix} 1 \\ 2 \\ 3 \\ 4 \\ 5 \\ 6 \\ 7 \\ 8 \\ 9 \\ 10 \end{bmatrix} y = \begin{bmatrix} 4868 \\ 5076 \\ 5236 \\ 5348 \\ 5404 \\ 5404 \\ 5348 \\ 5236 \\ 5076 \\ 4868 \end{bmatrix} \quad b = \begin{bmatrix} b1 \\ b2 \\ b3 \end{bmatrix} b1 = \sum_{x=1}^{n} x, b2 = \sum_{x=1}^{n} xy, b3 = \sum_{x=1}^{n} x^2 y \quad (15)$$

$$\begin{bmatrix} n & \sum_{x=1}^{n} x & \sum_{x=1}^{n} x^2 \\ \sum_{x=1}^{n} x & \sum_{x=1}^{n} x^2 & \sum_{x=1}^{n} x^3 \\ \sum_{x=1}^{n} x^2 & \sum_{x=1}^{n} x^3 & \sum_{x=1}^{n} x^4 \end{bmatrix} \times \begin{bmatrix} X_1 \\ X_2 \\ X_3 \end{bmatrix} = \begin{bmatrix} b1 \\ b2 \\ b3 \end{bmatrix} \quad (16)$$

$$X = bA^{-1}, \qquad X = \begin{bmatrix} 4597 \\ 294.66 \\ -26.78 \end{bmatrix}, \qquad a_i = X_i \tag{17}$$

$$y' = \frac{\Delta T}{\Delta t} = a_1 + a_2 x + a_3 x^2 \tag{18}$$

$$y' = skvGrplus = f_x = 4597 + 294.66x - 26.78x^2 \tag{19}$$

And by including it in MGSC, the future cash flow curve is defined Eq. (20) (**Figure 14**).

$$skvGr(x, T) = \lambda kv \cdot \int_0^x \frac{1}{(a \cdot T + b) \cdot \sqrt{2 \cdot \pi}} \cdot e^{\frac{-(x-\mu)^2}{kv \cdot (a \cdot T + b)}} dx + \left(4597 + 294.66x - 26.78x^2\right) dx \tag{20}$$

In the formula of the derivative curve, increasing Δt or dx, for example by half a month or 1 dx, increases the derivative of function 4597, from which ΔT is calculated, and thus MGSC is set at new intervals and with new values. So, for the given assumption of the extension of the project duration, the planned cost of the project ΔT increases by HRK 2.3 million, which increases the budget to HRK 28.3 million. How to assume the extension or reduction of the project can be realised by the step of differentiating the function or creating the intensity of resource use T′ = f (degree of rotation intensity of resource use). If we assume that ΔT is equal to the sum of the costs of all resource structures S_{ijk}, that is R_jEq. 21, then Δt can also be calculated from this data.

$$S_{ijk} = R_j = \sum_{i=1}^{nr} a_r i_r t_{ri} \tag{21}$$

In this way, we cover the planned normative money curve in all areas. In this way, the same is mapped equivalently to the valuation curve, which we define based on risk.

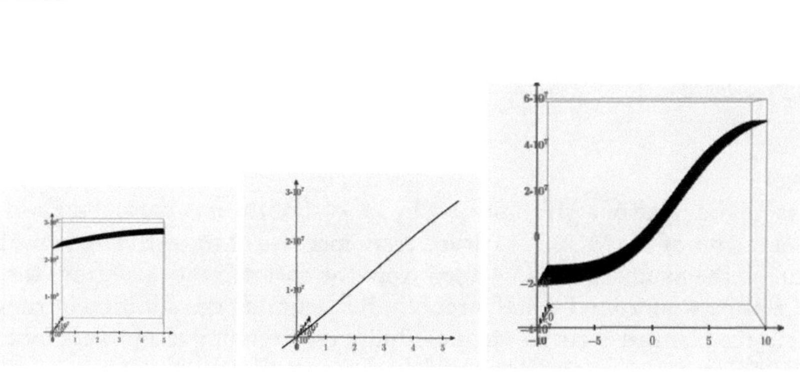

Figure 14.
MGSC project Zagreb-Zaprešić within 5,5 and –n to n months - adding half month.

5. Risk simulations with MGSC for production operations

Process business activities are highly susceptible to risks, and manufacturers are interested in business risks, which can be internal or external. Risks can be defined qualitatively and quantitatively, so such management is also defined. The analysis and control of risks, that is the identification and evaluation of risks through the regulation of the project, go through a cyclical process.

5.1 Risk management in processes

Japanese and American process control in corporate production processes has developed standards for product quality management, creating a general framework of standards for risk management published by various organisations in the world (ISO, IEC, etc.). In this way, the risk management process according to the standard ISO /IEC 31000 [25] was created, which covers all operations from the identification to the acceptance of risks with analysis and evaluation, as well as the assessment and management of risks through risk monitoring and consultation.

5.2 Quantitative risk analysis MGSK

To quantify the risk (Eg) of MGSK, mathematical, statistical and simulation modelling methods are used to simulate the financial situation of the project as it progresses. Based on previous research, trying to add the risk parameter (λr) to the investment level of 10% on the normative planned MGSC is highly probable, i.e. optimal for managing the investment, and the real cost curve with an estimate takes the form Eq. (22).

$$skvGr(x,T) = \lambda r \cdot \lambda kv \cdot \int_0^x \frac{1}{(a \cdot T + b) \cdot \sqrt{2 \cdot \pi}} \cdot e^{\frac{-(x-\mu)^2}{kv \cdot (a \cdot T + b)}} dx + \left(a1 + a2x - a3x^2\right) dx \tag{22}$$

Risk structuring by activities of the project plan in terms of the probability of occurrence of deviations from the planned cost variables of the project is done by the method of simple evaluation of site management. By dividing the MGSC of a given project into term tenths, the probability of the deviation or risk occurring in a given segment is obtained Eq. (23).

$$skvGr(x,T) = \lambda r \cdot \lambda kv \cdot \int_0^{x+\Delta t} \frac{1}{(a \cdot Tp + b) \cdot \sqrt{2 \cdot \pi}} \cdot e^{\frac{-(x-\mu)^2}{kv \cdot (a \cdot Tp + b)}} dx + \left(4597 + 294.66x - 26.78x^2\right) dx \tag{23}$$

Increasing the price of a given project by λr = 1.1 of the risk parameter added to the default parameter of the MGSC λkv leads to an increase of the estimated cost by about 2.6 million on the planning curve, which from the cost derivation defines the extension of the project duration by half a month. By assuming the addition of these cost increases to the planned S-curve, the possibility of extending the project by half a month is defined, bringing the budget of the estimation curve to 31 million (**Table 3**).

It is interesting to note that in practice estimators process bids and their manager usually reduces the estimate by 10% in order to be competitive.

ris	25,932	sum10	T′		2593	28,525.2
	ΔT	Δt	ΔT/Δt	Arctan(T′)	λR	ΔTr
1	2434	0.5	4868	1.570591	1.1	2677.4
2	2538	0.5	5076	1.570599	1.1	2791.8
3	2618	0.5	5236	1.570605	1.1	2879.8
4	2674	0.5	5348	1.570609	1.1	2941.4
5	2702	0.5	5404	1.570611	1.1	2972.2
6	2702	0.5	5404	1.570611	1.1	2972.2
7	2674	0.5	5348	1.570609	1.1	2941,4
8	2618	0.5	5236	1.570605	1.1	2879.8
9	2538	0.5	5076	1.570599	1.1	2791.8
10	2434	0.5	4868	1.570591	1.1	2677.4
11	2313	0.5	4597	1.570579	1.1	2544.3
	28,245	sum11				31,069.5

Table 3.
Planned and estimated (risk) iterations of the MGSC curve.

6. Automation and AI systems for process production (DSP, MGSC and CPM) MSP

Planning with the DSP code is achieved by using a combinatorial eq. [26] to define all paths in product manufacturing. Thus, the DSP code is a universal equation that can be used for all equations, including CPM equations or DSPiP modelling. By simulating the MSP model with a Gantt chart and an S-curve over the key resources given by the ABC curve, it comes down to defining the working time of the resources with costs (Eqs. 24–26) using the direction coefficient (a) of the linear direction of the representation of the resource costs plus the product with the intensity of the commitment of the resources (i) or the double direction coefficient.

$$\Delta \mathrm{t}(R_j) = \Delta \mathrm{T}(R_j)/T^{'} \tag{24}$$

$$\Delta \mathrm{t}_i(R_j) = \Delta \mathrm{T}\left(\sum_{i=1}^{nr} a_{ri} i_{ri} t_{ri}\right)/T_i^{'} \tag{25}$$

$$skvGr(x,T) = \lambda kv \cdot \int_0^x \frac{1}{(a \cdot T + b) \cdot \sqrt{2 \cdot \pi}} \cdot e^{\frac{-(x-\mu)^2}{kv \cdot (a \cdot T + b)}} dx + \Delta \mathrm{T}_i(R_j) \tag{26}$$

In the planned example itself, the curve ABC narrows the normative resource risk involved in about 80% of the project, and this is the Conject robot with a hydraulic pump involved in half of the project with almost 40% of this cost. By observing the given resource and reading the histogram of hydrodemolition activities, we simulate the method CPM, by increasing the resource. It can be seen that hydrodemolition takes 4 activities and 30 working days. If we double the labour intensity by

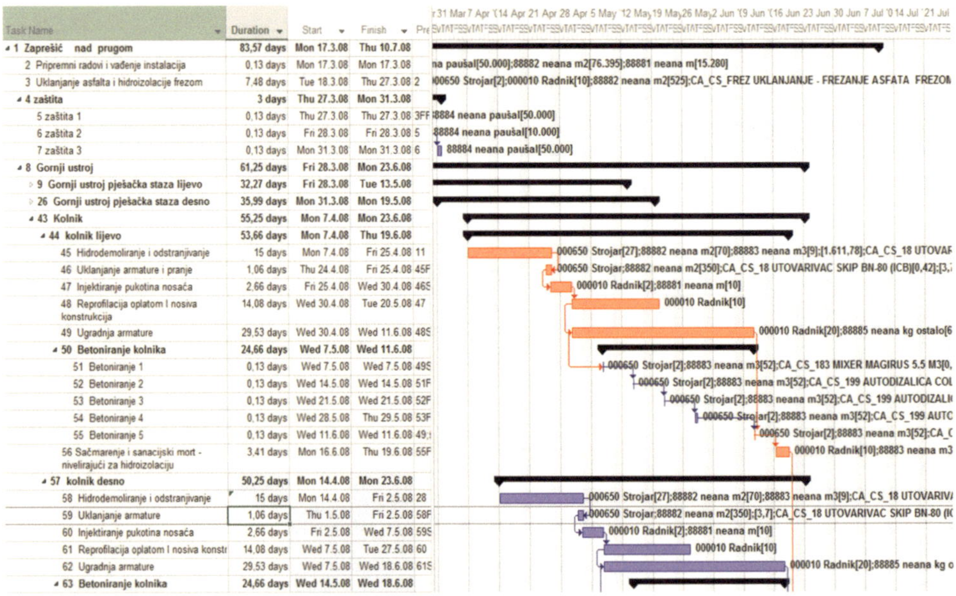

Figure 15.
Gantt chart of the revised Zaprešić project - introduction of a new work shift for the Conjet robot resource with hydraulic pump.

introducing a second shift, the duration decreases proportionally, (**Figure 15**), but we maintain the cost in relation to the total labour. But now the whole structure of MGSC DSP changes and the curve takes a different shape, however, the time does not decrease proportionally in the CPM method.

Instead of half a month, the reduction is 8 days, which is half a semester unit less in our MGSC system. So the budget in the normative project is slightly reduced compared to the budget of the original S-curve plan, and there is a greater density, that is an increase in monthly situations and a loss in risk costs. Another estimation method is possible by creating steps according to mathematical iteration methods (Eqs. 27–30) with the variables BAC and EAC.

$$T_{x+1} = T_x + \Delta T_{x+1} \tag{27}$$

$$T' = \frac{\Delta T}{\Delta x} = \frac{\Delta T}{\Delta t} = \frac{EAC - (T_{proc} - T_p)}{BAC} \tag{28}$$

$$\boldsymbol{T_{x+1} = T_x + \left(EAC - \left(T_{proc} - T_p\right)_{x+1}\right)} \tag{29}$$

$$\boldsymbol{T_{x+1} - T_x + \left(T_{proc} - T_p\right)_{x+1} = EAC} \tag{30}$$

By introducing the MGSC equation into the iteration series of the S-curve, the EAC query **Figure 16** was solved.

While simulating an increase in the BAC valuation curve with risk without extending the life of the project BAC costs the company about HRK 2.6 million, if BAC is extended by half a month, it is about HRK 5 million, which is a big warning from the EAC for management.

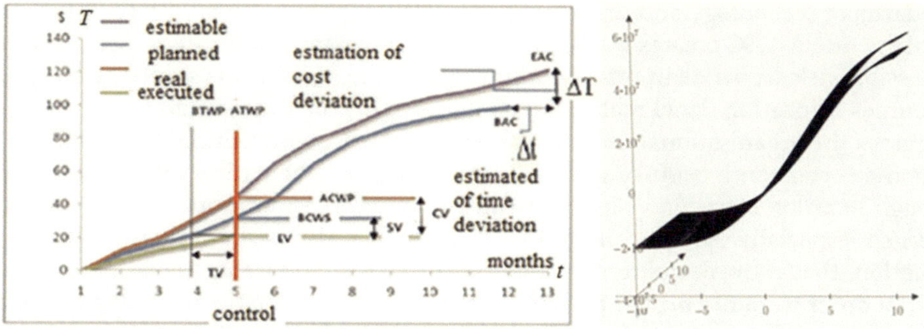

Figure 16.
Cash flow curve with risk in all states ($x_{estimat}$ = 5.5, x_{plan} = 5).

7. Conclusions

Mathematical modelling of the organisation is catching up with technology, and with the advent of computer and software support, the desire to define a dynamic POB or PDM is being realised. This is achieved through model standardisation with vector and parametric normalisation of production processes, and mathematical and statistical methods are used to approach precision and reduce risk or increase sustainability of projects. In order to define the precision of finances in production processes, a new attitude in defining the model is needed, without rewriting old technologies into software technologies, in addition to the technological support provided by ITC. Moreover, every production wants to be improved, that is to achieve more indicators of the economic principles of efficiency, economy and productivity. Thus, in the process industry, the direct cost method is developed and the cash flow curve is presented in the project presentation of the budget. By creating a cash flow curve by defining the planned MGSC based on the case study method and an estimate based on some data from practice, that is from the data of the Međimurje Graditeljstvo company, an overall model of the cash flow curve is created without realisation. Thus, defining the general equation of the S-curve in multidimensional space, the entire series of S-curves is obtained as the supremum and infimum of the S-curve data set in a project. This creates a probability domain using statistical curves so that better risk management in the project is possible, that is production management is also possible with an assumed modified Gaussian S-curve, which opens a new research area for BAC and EAC dimensions in the project. It is recommended that further research be conducted based on implementation and planning projects to determine the risk coefficient or λr as realistically as possible and to seek a creative team leader in corporate security who has the skills of an IT manager [27] who has a new tool for defining risks in a corporate project. The iterative structural DSP method enables further linkage of product surveillance with the ITC system and leads to fragmentation, that is automation and robotisation of large infrastructural investments in a circular sustainable economy in the economic sense with an increase in the principles of economy, productivity and profitability. In the humanistic sense, AI systems free workers from dangerous and difficult work and open up DSP and cost-effective constant modelling and simulation of model standardisation. This, in turn, strives for the sustainability of the business model with AI [28] and cyberphysical processes (CPS), that is a system for initiating new discoveries in defining the development of standards of technology and organisation in MP. This connection with the sequence numbers on RMGSC creates a simulation of the development of the project according to possible risk scenarios in relation to the cost variable, which is similar to the Joint Resources Allocation Panel method that CPS uses for the time variable. The team also addresses the organisational differential, which has been introduced into engineering since the seventeenth century with the invention of mathematical differential models through iteration equations, and in the mid-twentieth century, operational research emerged, especially dynamic programming, which is also based on mathematical induction, that is the definition of iteration equations favourable to computers. However, in order to round off the cycle for simulation or automatic modelling of the S-curve, the equations need to be refined, especially in the case of the 4D model, where the resource ABC is considered as the third dimension in the function of s t and T, with the possibility of daily project management with differentiation of the intensity of the resources in the project, especially with the method DSP or DSP code similar to XML code. The management team becomes dynamic, a digital twin system is created and in the organisation, we simulate the reality of the projects through the model standardisation recommended by the DSP method.

Conflict of interest

The authors declare no conflict of interest.

Notes/thanks/other declarations

Thanks for IntechOpen.

Author details

Vladimir Križaić
Politechnic of Medjimurje, Čakovec, Croatia

*Address all correspondence to: vladimir.krizaic@gmail.com

References

[1] Orešković M. Monitoring graditeljskog projekta. FGZ-Građevinar. 2019;**71**(11):965-973

[2] Lončarić R. Organizacija izvedbe graditeljskih projekata. Zagreb: Sveučilišna tiskara; 1995

[3] Huang C, Fisher N. An alternative methodology for selecting enterprise resource planning (ERP) systems in a Typical Architecture/Engineering/ Construction (AEC) Company. Interna tional Journal of Construction Manage ment. 2014;**5**(1):39-57

[4] Marinković D, Stojadinović Z, Ivanišević N. Dinamički planovi utemeljeni na ciklusima rada. Građevinar;**65**:993-1002

[5] Križaić V. Application of norms models with vectoral system in construkction projects. Journal of Civil Engineering an Architecture. USA. Jun 2014;8(6):722-728. ISBN 1934-7359 str. 728-728

[6] Abdelhady MI, Abdelgadir AK, Al-Araimi F, AL-Amri K. Using algorithm in parametric design as an approach to inspire nature in architectural design. ICIVC 2021, PALO 15. Springer Nature. 2022. pp. 1-18. DOI: 10.1007/978-3-030-97196-0_10

[7] Križaić V. Automation of Construction Propduction Using the Dsp Method, CCC 2023. Keszthely, Hungaria: Procedia Engineering; 2023

[8] Zangh S, Zangh L, Wang D, Zhou B, Le Z. Research on the Stability of Urban Bus Network Based on Complex Networks Theory, Hilbert Space. China Mainland, Hubei: ASCE; 2019. DOI: 10.1061/9780784482292.154

[9] Guler H. Optimizacija održavanja i remonta željezničkih kolosijeka primjenom genetičkih algoritama. FGZ – Građevinar. 2016;**68**(12):979-993

[10] Faller C, Höftmann M. Service-oriented communication model for cyber physical-production-systems. In: 11th CIRP Conference on Intelligent Computation in Manufacturing Engineering, Procedia CIRP. Vol. 67. Elsevier BV. 2018. pp. 156-161

[11] Drljača M. Oblikovanje modela poslovnog upravljanja u skladu s modelima TQM. In: Kvalitetom Protiv Recesije. Zagreb & Rovinj: Hrvatsko društvo menadžera kvalitete; 2013

[12] Hübner F, Volk R, Schultmann F. Project management standards: Strategic success factor for projects. International Journal of Management Practice. 2018; **11**(4):372

[13] Garcia de Soto B. Devenloping a Digital Twin on a University Campus to Support Efficient and Sustainable Buildings, CCC 2023. Keszthely, Hungaria: Procedia Engineering; 2023

[14] Križaić V, Hranj D. Daily monitoring by IIS. In: 11th OTMC. Dubrovnik: Croatian Association for Construction Management; 2013. ISBN 978-953-7686-03-1, str. 26

[15] Sikavica P. Organizacija. Zagreb: Školska knjiga; 2011

[16] Andrijanić I, Gregurek M, Merkaš Z. Upravljanje poslovnim rizicima. Zagreb: Liberta-Plejada; 2016

[17] Križaić V. Organization Limit – Modeling and Simulation, IC Investmen Strategies and Management of Construction. Brijuni: Society of

Croatian Construction Managers; 1994 ISBN 953-96245-0-9, str. 345-350

[18] Križaić V, Rodiger T, Buč S. Informa tizirani menadžer u vektorskoj organizac iji. Journal of Civil Engineering and Architecture. 2021;**15**(8):419-428. DOI: 10.17265/1934-7359/2021.08.003

[19] Radujković M, Car-Pušić D, Ostojić Škomrlj N, Vukomanović M, Burcar Dunović I, Delić D, Meštrović H. Planiranje i kontrola projekata. Građevinski fakultet Sveučilišta u Zagrebu; 2012

[20] Križaić V. Application of norms models with Vectoral system in Construkction projects. Journal of Civil Engineering and Architecture. 2014; **8**(6):722-728

[21] Križaić V, Hranj D. Project Planning by Modified Gauss S-Curve. Opatija, Croatia: CCC 2020; 2020. DOI: 10.3311/CCC2020-047

[22] Križaić V, Rodiger T, Baksa S. Simulation project management by modified gaussian S-curve. In: 14th WCCM-ECCOMAS Congress, January 2021. 2021. DOI: 10.23967/wccm-eccomas.2020.309

[23] Zečević A. Matematika za informatičare. Beograd: Naučna knjiga; 1986

[24] Mladenović N, Spasić V, Jovanović M. Numerički metodi za mikroračunare. Beograd: Tehnička knjiga; 1986

[25] Krakar Z. Korporativna informacijska sigurnost. Varaždin: FOI; 2014

[26] Križaić V. Planning through combinatorics. In: VII International Conference "Organization, Technology and Management in Construction". Zadar: Croatian Association for Organization in Construction; 2006

[27] Šarčević M. Komunikacijska uloga i kompetencije IT menadžera. Zagreb: Filozofski fakultet; 2016

[28] Shan Y. Virtual reality in China: Is there a sustainable business model for virtual reality content enterprises. Cultural Science Journal. 2019;**11**(1): 54-67

Chapter 6

Warehouse Operations: An Examination of Traditional and Automated Approaches in Supply Chain Management

Olumide F. Odeyinka and Olumide G. Omoegun

Abstract

This paper explores the different approaches to warehouse operations within a supply chain context. It highlights the significance of warehouses in efficiently and effectively meeting customer demands while minimizing costs. The study examines various types of warehouses and analyzes the activities involved in warehouse operations. Furthermore, this paper discusses the challenges and benefits of warehouse operations in developed and developing countries. It emphasizes the need for efficient warehouse management to enhance customer satisfaction, reduce costs, and improve overall productivity. The integration of warehouses into the supply chain process is also emphasized, emphasizing their critical role in facilitating the flow of goods. This paper introduces the Warehouse Management System (WMS) concept as a technological solution for controlling the inflow and outflow of goods. It delves into WMS functions and highlights its importance in inventory control, order management, and system reliability. Additionally, this paper examines the differences between traditional manual warehousing and automated approaches, taking into account factors such as cost, efficiency, and suitability for different types of organizations. In conclusion, the study suggests that automated warehousing is more suitable for larger companies operating in countries with limited manual labor availability. On the other hand, traditional warehousing may be more applicable in developing countries with higher unemployment rates.

Keywords: warehouse operations, cost minimization, technological solutions, supply chain efficiency, warehouse management system, inventory control, customer satisfaction

1. Introduction

The growing interconnectedness of nations, the progressive evolution of the global economy, and the rise of consumer-driven industrialism have fostered a heightened need for logistics, transportation, and warehousing solutions. Within the supply chain, warehouses play a pivotal role by substantially facilitating the storage of goods from their production phase until they are delivered to consumers upon request [1]. A warehouse is not just a simple room or space; it is a purpose-built facility created by suppliers or manufacturers to store goods quickly before they are supplied to customers. According to Kolinski and Sliwczynski [2], in the fiercely competitive landscape of modern business, a warehouse serves not only as a mere storage facility for inventory but also as a hub for the efficient execution of value-added services. Its primary purpose, as defined by suppliers or manufacturers, is to temporarily store goods until they are ready to be delivered to customers. This setup enables businesses to promptly fulfill customer demands as they arise. The fundamental objective of warehousing is to enhance the efficient and effective flow of goods from sellers to buyers, ultimately eliminating additional costs that may be incurred when urgent needs for goods arise on the supplier's side [3]. Furthermore, warehouses give suppliers the advantage of making large-volume purchases, resulting in cost savings and enhanced economies of scale [4]. Warehousing addresses the dichotomy between the inflow and outflow of goods in organizations that adopt it. This is achieved through the coordinated efforts of various components such as storage racks, material handlers, warehouse workers, and specialists. These elements work together to ensure the smooth and efficient functioning of the warehouse, allowing for the timely receipt, storage, and distribution of goods.

Goods stored in a warehouse have not yet reached their final destination. However, they are subsequently distributed optimally from the warehouse itself, which can also be referred to as a distribution house [5]. This facility is strategically designed to accept new orders, implement them efficiently, and ensure the timely distribution of goods through the appropriate channels. In order to keep up with evolving industry trends, modern warehousing practices have embraced new technologies that enhance distribution processes and enable effective management of available resources. Historically, warehousing has primarily been perceived as minimizing costs without necessarily adding value to existing products. However, it also serves the critical function of preserving the original quality of goods [6]. The warehouse now plays an integral role in enhancing the movement of goods from one location to another, effectively functioning as a dispensation supply chain [7, 8].

Some businesses previously viewed Distribution Control Systems (DCS) and warehouses as standardized supply chain components solely focused on cost management. However, there is now an increasing recognition that warehouses have transformed into indispensable elements rather than mere cost centers [9, 10]. This shift in perspective is motivated by several factors, such as heightened market volatility, capacity limitations, evolving regulations, substantial shifts in consumer demographics and purchasing behavior, and stricter requirements imposed by customers and suppliers [11]. With the rapid advancements in technology, warehouses possess the capacity to foster competitive differentiation and facilitate profitable growth for enterprises [12]. A comprehensive review of the literature on smart warehouse operations management was carried out by Zhen and Li [13]. The authors identified several research gaps and challenges. They discovered that the previous studies on smart warehouses were fragmented and lacked a comprehensive perspective. They also observed a dearth of empirical studies assessing the effectiveness and efficiency of smart warehouse technologies. The researchers recommended that future research should concentrate on developing a unified framework for managing smart warehouse operations and conducting more empirical studies to evaluate the performance of such technologies. A recent study conducted by Kembro and Norrman [14]

DOI: http://dx.doi.org/10.5772/intechopen.113147

examined the transition from manual to smart warehousing, specifically focusing on Swedish retailers. This study aimed to conceptualize and operationalize the concept of smart warehousing. The findings revealed that future smart warehouses will possess characteristics such as automation, autonomy, digitization, and connectivity. The research paper established two dimensions to operationalize smart warehousing: the level of automation and the level of digitalization and connectivity of information platforms. Building upon these insights, the study presented 16 theoretical propositions that elucidate various pathways for retailers to implement smart warehousing, considering contextual factors. The empirical findings and theoretical discussions offer valuable practical guidance, including trends and considerations for the selection and benchmarking of automation and complementary technologies in warehouse operations.

Based on the above-mentioned, this study aims to examine various types of warehouses and analyzes the activities involved in warehouse operations. Furthermore, this paper discusses the challenges and benefits of warehouse operations in developed and developing countries. The integration of warehouses into the supply chain process is also emphasized, emphasizing their critical role in facilitating the flow of goods. This paper delves into Warehouse Management System (WMS) functions. It highlights their importance in inventory control, order management, and system reliability. Additionally, this paper examines the differences between traditional manual warehousing and automated approaches, considering factors such as cost, efficiency, and suitability for different types of organizations.

2. Warehouse operations

Organizations typically establish three primary categories of warehouses: raw materials, semi-finished goods, and finished goods. Depending on the nature of their business, companies may also create other types of warehouses. When categorizing warehouses, common features are considered. Raw materials storage warehouses hold capital goods used in producing other items, specifically those needed by organizations, especially manufacturing companies, for their intended products [15]. Semi-finished product storage warehouses house goods in progress before they become consumer goods, serving purposes such as branding or acting as assembly points for multi-stage goods [16]. Finished goods storage warehouses store products ready for final consumers, serving as destinations for producers, wholesalers, retailers, and customers to collect their orders [17]. This storage is critical for companies to meet expected demand, prevent shortages from unforeseen orders, and store new products effectively.

Besides receiving new products, checking their quality, and accepting them, warehouses perform other activities. They play a crucial role in resource sharing among different departments, promoting efficient resource utilization and coordination. Additionally, warehouses provide secure storage for products, protecting them from damage caused by environmental factors and potential hazards [18]. They organize and arrange company products to optimize space and ensure efficient inventory management, making it easier to access and retrieve items when fulfilling orders. Warehouses also prepare products for shipment by packaging them appropriately and verifying order details to ensure accuracy. Overall, warehouses serve as integral components of effective operations, facilitating resource sharing, secure storage, and streamlined order preparation for different categories of goods.

Warehouse operations encompass the daily tasks involved in storing and transporting goods within a warehouse facility. These tasks comprise activities such as receiving, managing inventory, selecting orders, packing items, arranging shipments, and overseeing distribution. The primary objective of warehouse operations is to effectively utilize space, equipment, and labor while meeting the needs and demands of customers [19]. **Figure 1** illustrates a diagram outlining warehouse design and operation planning classification. Designing and operating a warehouse involves addressing numerous challenges to fulfill specific requirements. These challenges include allocating resources like space, labor, and equipment to various warehouse functions. Each function must be implemented, operated, and coordinated effectively to meet system requirements, such as capacity, throughput, and service while minimizing resource costs.

Similarly in Saleheen, Miraz, Habib, and Hanafi [21], the challenges of warehouse operations were investigated. The work presented a case study focused on a retail

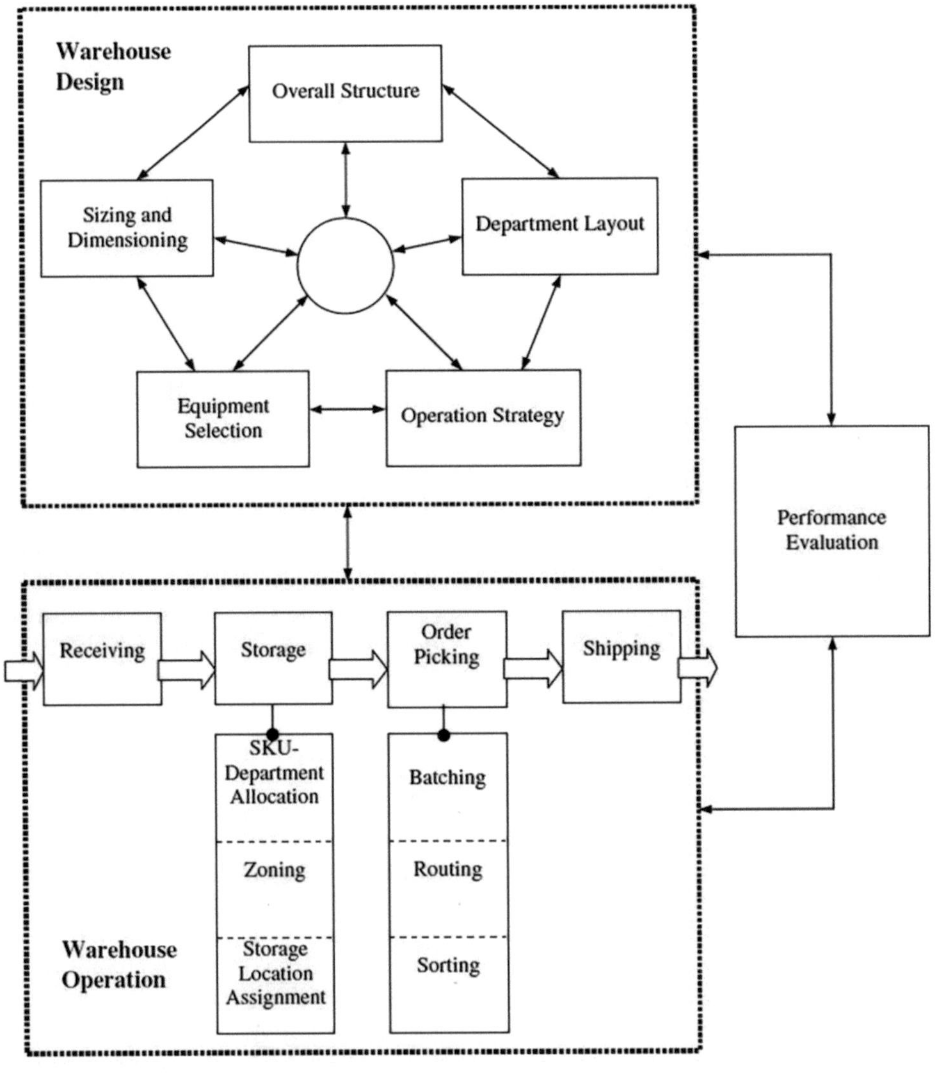

Figure 1.
Classification diagram of warehouse design and operation planning [20].

supermarket in Bangladesh, examining the complexities of warehouse management practices and highlighting the impact of retail supply chain dynamics on the role of the warehouse. The study identified external factors, market trends, and potential barriers that influenced the overall performance of warehouse operations. Additionally, the paper identified opportunities to enhance warehouse performance in terms of planning, design, and operation. The authors offered recommendations to industry practitioners to achieve higher productivity levels in warehouse management.

Also, Gunasekaran, Marri, and Menci [22] focused on enhancing the effectiveness of warehousing operations. Authors developed a conceptual framework based on four main factors: inventory management, material handling, layout design, and information systems. This framework was then applied to a case study involving a warehouse responsible for managing spare parts for an automobile company. The authors collected data on the current state of the warehouse, identifying areas of concern such as excessive inventory levels, underutilized space, inefficient material handling, and inadequate information systems. To address these issues, the authors proposed solutions aimed at improving warehousing operations. These solutions included reducing inventory levels, redesigning the layout, implementing an automated storage and retrieval system (AS/RS), and adopting a WMS. The proposed solutions were evaluated through simulation and cost-benefit analysis, which revealed significant improvements. The results showed a 60% reduction in throughput time, a 40% increase in space utilization, and a 30% cost savings. In conclusion, the authors emphasized that the conceptual framework had the potential to assist warehouse managers in enhancing the effectiveness of their operations, ultimately leading to increased customer satisfaction.

Warehouse operations in both developed and developing countries offer various benefits and face specific challenges. The benefits of warehouse operations include the availability of goods on demand, reduced costs associated with emergency orders and unforeseen consequences, the promotion of well-structured storage spaces, and the facilitation of staff productivity through clear production and sales targets. However, there are challenges to overcome in ensuring successful warehouse operations. In some cases, the lack of necessary processes, staff, and programs can hinder efficiency, particularly in developing countries where resources may be scarce or expensive to acquire. Repetitive processes can lead to monotony in developing countries, while system failures may interrupt operations in technologically advanced developed countries. Accurately arranging goods for easy accessibility can be challenging, and the requirement for sufficient warehouse space may lead to congestion in both developed and developing countries. Furthermore, accessing specific goods within the warehouse may become time-consuming when maintaining designated spaces [14, 23, 24].

3. Warehouse management system (WMS)

A warehouse management system (WMS) is a software application or platform created to enhance and streamline the efficiency of warehouse operations. It equips businesses with the necessary tools to efficiently handle and monitor inventory, manage warehouse tasks, and enhance overall supply chain processes [25]. Khan, Huda, and Zaman [26] describes a WMS as a system that utilizes advanced technology (comprising both hardware and software components) to regulate the movement of goods throughout various stages, starting from the moment an order is placed, continuing through acquisition, quality inspection, processing, storage within the warehouse, and finally, dispatching them as per demand. It is also described as a database-driven computer application that improves the warehouse's efficiency by directing cutaways and maintaining accurate inventory by recording warehouse transactions [27]. The systems also directed and optimized stock based on real-time information about the status of bin utilization. They often utilized Auto ID Data Capture (AIDC) technology, such as barcode scanners, mobile computers, wireless LANs (Local Area Networks), and potentially Radio-frequency identification (RFID) to monitor the flow of products efficiently. Once data had been collected, there was either batch synchronization or a real-time wireless transmission to a central database. The database could then provide useful reports about the status of goods in the warehouse.

Warehouse Management System (WMS) oversees the production and supply chain processes, ensuring a continuous flow of materials for production, minimizing the risk of stockouts, and facilitating the transformation of raw materials into finished products. The main objective is to enable efficient and effective transportation of goods from the warehouse to their intended destinations [28]. In addition to inventory control, order acceptance, and order response, WMS performs various other functions in addition to its primary roles. It assesses the need for expanding company capacity and evaluates the potential benefits of adopting advanced technology. This assessment takes into account factors such as increased efficiency, cost savings, and improved productivity. It also considers the long-term sustainability of the chosen technology, ensuring that it can support the organization's growth and evolving needs [29]. WMS plays a crucial role in evaluating the reliability of the current system. It examines the existing warehouse management processes and identifies any outdated systems or practices that may hinder operational efficiency. When necessary, WMS helps implement necessary changes and upgrades to ensure that the system remains up-to-date and optimally functional [30]. Another significant function of WMS is analyzing warehouse capacity requirements. It considers factors such as market size, demand fluctuations, and available funds to determine the appropriate capacity needed to efficiently store and manage inventory [31]. WMS also provides flexibility for organizations in managing their warehouses. It allows companies to choose between managing their warehouse in-house or opting for larger warehouses. This flexibility is valuable for businesses as it accommodates their specific needs and growth plans. For instance, companies can start small and gradually expand their warehouse operations as required, thanks to the scalability offered by WMS [32, 33].

There are two primary types of WMS: on-premise and cloud-based. On-premise WMS refers to software hosted and maintained on-site using the warehouse's own servers, while cloud-based WMS is hosted online in the cloud and accessible from any connected device. Cloud-based WMS offers advantages in terms of scalability, flexibility, and automatic updates compared to on-premise WMS [34, 35].

WMS software can integrate with various internal and external software systems, including ERP, OMS, TMS, barcode scanning, accounting, and e-commerce platforms. This integration enables the automation of billing, inventory management, receiving, shipping, and picking and packing through mobile barcode scanning. Distinctions can be made between WMS and OMS (Order Management System). While WMS software captures, records, and analyzes data on all day-to-day functions performed within a warehouse, OMS software specifically focuses on monitoring the movement of inventory and orders throughout the supply chain. While OMS do not replace WMS, it can complement it within a warehouse technology stack [36–38].

WMS software is valuable for both private and 3PL (third-party logistics) warehouses that handle various order fulfillment types, including pallet in/pallet out, B2B, B2C, e-commerce, and omnichannel. It caters to the needs of warehouses serving multiple industries such as retail, apparel, bulk goods, raw materials, cold storage, nutraceuticals & pharmaceuticals, wine & spirits, and hazardous materials [39–41].

4. Warehousing systems

There are two different warehousing systems, traditional and automated warehousing.

4.1 Traditional warehousing

Traditional warehousing refers to the conventional and established methods of storing and managing inventory in a warehouse facility. Kamali [42] stated that traditional warehouses are typically designed to effectively utilize a limited amount of space due to specific limitations or constraints while simultaneously meeting all the essential requirements for day-to-day operations within the facility. They typically involve manual processes and equipment without extensive automation or advanced technology. Richards [10] suggested that traditional warehousing is a manual system in which staff within an organization manually handle order picking and deployment and brand and rebrand the company's products. Although some machinery may be involved, it is operated solely by humans. Despite the repetition and increased labor costs, growing companies still adopt this warehousing system regardless of the number of orders they receive. However, this approach poses risks for the employees. In developing countries, companies with products that experience seasonal demand also employ this method. They hire additional labor during peak periods while retaining their core staff [43]. Nonetheless, the use of machinery can increase operational costs, regardless of whether the machine is active or inactive. Traditional warehouses are typically organized using standardized storage systems such as pallet racking, shelving, and bins. The layout is designed to accommodate manual movement of goods and optimize space utilization [44]. Material handling in traditional warehousing relies heavily on manual labor. Workers use equipment such as forklifts, pallet jacks, and hand carts to move goods within the warehouse. Loading and unloading of trucks and containers are also done manually [45].

Van Geest, Tekinerdogan, and Catal [46, 47] mentioned that inventory management in traditional warehousing is often performed through manual tracking and recording systems. Paper-based or basic computer systems may record inventory levels, receipts, and shipments. Order picking in traditional warehouses is typically done manually. Workers physically locate and retrieve items from their storage locations based on paper pick lists or basic technology aids like handheld scanners [46–48]. While traditional warehouses may use basic technology for inventory management and order processing, they often lack advanced automation and integration capabilities. WMS may be utilized for basic inventory tracking and control. Traditional warehousing may have minimal automation, especially when compared to modern, highly automated facilities. Manual processes and equipment are the primary means of handling goods, with little or no use of automated storage and retrieval systems (AS/RS), conveyor systems, or robotics. Traditional warehousing may also offer limited flexibility and scalability due to the reliance on manual labor and fixed storage systems. Adapting to changing operational needs or accommodating growth may require physical layout modifications and additional labor. While traditional warehousing may lack some of the efficiency and speed advantages of more modern approaches, it can still be effective for certain businesses with lower throughput volumes or unique product requirements. However, many warehouses are now adopting elements of modern warehousing, such as incorporating automation, advanced technology, and data-driven optimization, to improve operational efficiency and meet evolving customer demands [47].

Kamali [42] used a Bahraini company as a case study to illustrate the characteristics and disadvantages of traditional warehousing. The paper described the layout as shown in **Figure 2** and the design of the traditional warehouse, which consists of different areas for loading, unloading, receiving, storing, picking, assembling, dispatching, and value-added logistics. Kamali [42] identified some disadvantages of traditional warehousings, such as high material handling costs, inefficient operations and workflow bottlenecks, negligence and damaged materials, inefficient space management, and inefficient materials handling equipment. These problems could reduce the revenues and profitability of the company in the long term. Dukic and Opetuk [49] reviewed the literature on the optimal layout design of a traditional warehouse, which aims to minimize the total warehousing costs, such as material handling, storage, and fixed costs. Their study classified the layout design problem into two categories: single-block and multi-block layout design. The single-block layout design problem is considered a rectangular storage area with one or more aisles. The multi-block layout design problem considered a storage area divided into several blocks, each with its own aisles and cross-aisles. They also discussed various models' assumptions, objectives, constraints, and solution methods for single-block and multi-block layout design. The authors identified some limitations and challenges of traditional warehouse layout design, such as the difficulty of incorporating stochastic factors, the complexity of solving large-scale problems, and the lack of integration with other warehouse functions.

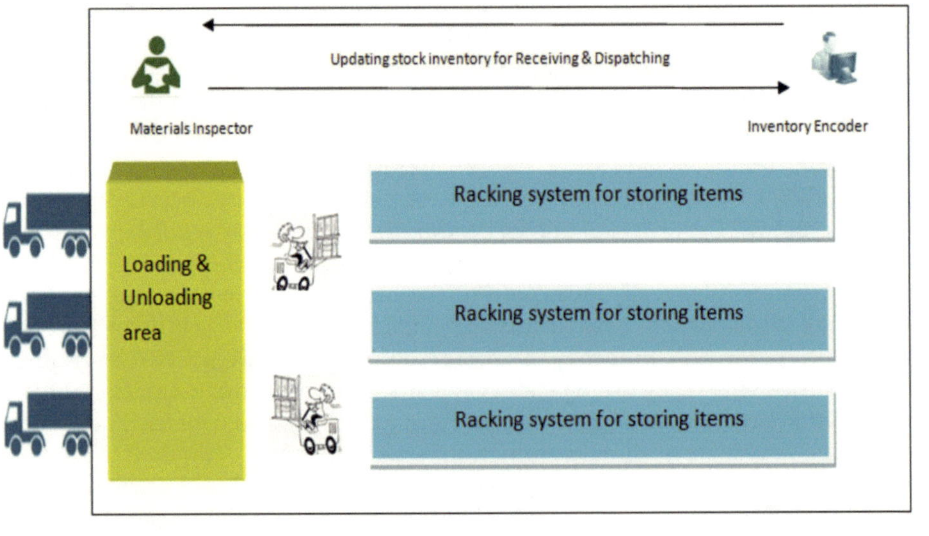

Figure 2.
The layout of a traditional warehouse [42].

4.2 Automated warehousing

Automated warehousing encompasses the utilization of advanced technology and automation systems to enhance the efficiency and optimization of warehouse operations. It involves integrating diverse automated equipment, robotics, and software systems to improve productivity, accuracy, and overall effectiveness [50]. Key components of automated warehousing include AS/RS. These systems comprise tall racking structures equipped with automated cranes or robotic systems that retrieve and store goods from specified locations. AS/RS efficiently handles substantial inventory volumes and enables high-density storage [10]. Automated conveyor systems are used to transport goods within the warehouse. They can move items between workstations, sort and merge products, and connect processing areas, such as picking, packing, and shipping zones. Conveyor systems enhance material flow and reduce manual handling [51]. Robots and Robotics and Automated Guided Vehicles (AGVs) are used for tasks such as picking, sorting, and palletizing goods. These automated systems can operate autonomously or be guided by sensors, cameras, or magnetic strips on the warehouse floor. They can perform repetitive tasks quickly and accurately, reducing labor requirements and improving productivity [52, 53].

WMS software is crucial in automated warehousing to manage and control inventory, orders, and workflows [54]. It integrates with various automation systems and provides real-time inventory levels, order status, and operational performance data. WMS optimizes processes, improves inventory accuracy, and enables efficient order fulfillment [55]. Pick-to-Light and Put-to-Light Systems use visual cues, such as lights or displays, to guide warehouse operators through the picking and putting processes [10]. They help reduce errors, increase picking speed, and enhance order accuracy. Automatic Identification and Data Capture (AIDC) technologies, such as barcode scanners, RFID (Radio Frequency Identification) systems, and voice recognition devices, automate data collection and capture [56]. They enable accurate inventory tracking, reduce manual data entry, and enhance order visibility. Automated warehouses may incorporate energy-efficient technologies, such as LED lighting, automated climate control systems, and smart energy management solutions [57]. Sustainability practices, such as recycling and waste reduction, can also be integrated. Automated warehousing offers benefits such as increased throughput, improved order accuracy, reduced labor costs, and enhanced operational efficiency [58]. However, implementing automation requires careful planning, significant investment, and consideration of factors like product characteristics, order profiles, and ROI calculations. Additionally, proper training and maintenance are essential to ensure the smooth operation of automated systems.

4.3 Discussion

Various studies have outlined the outlined different research imperatives. For example, Zhen and Li [13] discussed the challenges and opportunities associated with smart warehouse operations management and proposed potential areas for future research. These areas included performance evaluation, coordination and collaboration, human factors, and social responsibility. The authors concluded that smart warehouse operations management held significant promise as a research field that could make valuable contributions to the advancement of logistics and supply chain management. Similarly, Kamali [42] outlined the benefits of automated warehouses, such as better warehouse optimization, reduced labor costs, improved accuracy and reliability, enhanced customer service, and faster shipping. However, it also identified some challenges and limitations of smart warehouses, such as high initial costs, complex integration, skill requirements, maintenance issues, and software glitches. The research suggested some possible solutions and recommendations to overcome these challenges and maximize the potential of smart warehouses, such as WMS, Artificial Intelligence (AI), cloud computing, robotics, and blockchain. In the study "Robotized and Automated Warehouse Systems: Review and Recent Developments", Azadeh, De Koster, and Roy [59] reviewed new categories of automated and robotized handling systems, such as shuttle-based storage and retrieval systems, shuttle-based compact storage systems, and robotic mobile fulfillment systems. The research revealed that despite the growing utilization of various robotic systems and applications, there was a lack of academic literature comprehensively investigating them. These systems possessed unique characteristics, such as autonomous control and dynamic operation within a networked environment. As a result, novel models and approaches were needed to address the design and operational control challenges associated with these systems, particularly in terms of integrating subsystems. The future trend in warehousing was expected to involve integrated robotized warehouse systems, which would require a reassessment of crucial aspects of warehouse design, planning, and control. This entailed revisiting methodologies for designing layouts, selecting storage and order picking systems, determining storage slotting strategies, optimizing order batching, defining picker routing paths, and assigning pickers to orders within the context of the new robotized warehouse paradigm. Custodio and Machado [60] did a review and an innovative framework for a flexible automated warehouse. The researchers reviewed published papers in the last 10 years that were related to flexible automation in warehouses. They constructed a framework that could guide future researchers in developing an innovative conceptual model for application in future warehouses. They concluded that the crucial aspect of achieving a flexible automated warehouse was the integration of automated equipment, data collection technologies, and management solutions.

5. Conclusion and recommendations

5.1 Conclusion

This work highlights the crucial role of warehouses in the global supply chain. It has been posited that warehouses are no longer just storage facilities but essential components for efficient value-added services and meeting customer demands. They play a significant role in storing, managing, and distributing goods, contributing to the smooth flow of products and eliminating unnecessary costs. They have become integrated into the supply chain, embracing modern technologies for enhanced distribution processes and resource management. Warehouse operations involve various activities such as inventory management, order selection, packing, shipping, and distribution. Challenges in warehouse operations include resource allocation, market trends, inventory management, layout design, and information systems. Overcoming these challenges is crucial for higher productivity and customer satisfaction. Effective warehouse operations are vital for companies to meet customer demands, optimize space utilization, and ensure timely distribution. WMS are software applications that streamline warehouse operations, including inventory management and task management. They evaluate the current system's reliability, identify outdated practices, and implement necessary changes. WMS also helps determine the appropriate capacity needed for efficient inventory management.

Warehouse management can be accomplished through traditional methods or modern systems that employ automation to control the flow of goods. Traditional warehousing systems rely primarily on manual processes for order picking and deployment. They heavily rely on human labor and retain only essential staff members, hiring additional labor during peak periods. While traditional methods are user-friendly, they are prone to human errors. On the other hand, automated warehouse operations involve the use of technology to reduce manual work and streamline processes such as ordering, inventory control, and storage. Although automation brings operational ease, the installation and maintenance costs are high, making it feasible mainly for larger organizations. More so, automated warehousing is better suited for large companies, particularly in countries where manual labor for warehouse operations is scarce and expensive to hire. Traditional warehouse operations find more applicability in developing countries grappling with high unemployment rates. They are advantageous for small companies that cannot afford the expenses associated with deploying modern automation technology in their warehouses.

Potential areas of further research have also been outlined from the different literatures reviewed. These areas included performance evaluation, coordination and collaboration, human factors, and social responsibility of smart warehouse management systems, integration of artificial intelligence and robotics in warehouse systems, extension of design of warehouse systems and many more.

5.2 Recommendation

Based on the findings of this study, the following recommendations have been made:

- Businesses should consider implementing a WMS to enhance the efficiency of their warehouse operations and improve overall supply chain processes.

- When choosing a WMS, organizations should evaluate their specific needs, scalability requirements, and integration capabilities with other software systems.

- Regular evaluation and assessment of the warehouse management system should be conducted to identify areas for improvement and implement necessary changes or upgrades.

- Organizations should assess the costs, benefits, and feasibility of implementing automated warehousing systems, considering factors such as product characteristics, order profiles, and return on investment calculations.

- Training and maintenance programs should be established to ensure the smooth operation of the implemented WMS and automated warehousing systems.

Author details

Olumide F. Odeyinka[1*] and Olumide G. Omoegun[2]

1 Department of Systems Engineering, University of Lagos, Lagos, Nigeria

2 Department of Engineering and Mathematics, Sheffield Hallam University, United Kingdom

*Address all correspondence to: oodeyinka@unilag.edu.ng

References

[1] Karim NH, Abdul Rahman NSF, Md Hanafiah R, Abdul Hamid S, Ismail A, Abd Kader AS, et al. Revising the warehouse productivity measurement indicators: Ratio-based benchmark. Maritime Business Review. 2021;**6**(1):49-71. DOI: 10.1108/MABR-03-2020-0018

[2] Kolinski A, Sliwczynski B. Evaluation problem and assessment method of warehouse process efficiency. In: Business Logistics in Modern Management. 2015: Proceedings of The 15th International Scientific Conference Business Logistics in Modern Management. Osijek, Croatia; 2015

[3] Balcik B, Beamon BM, Krejci CC, Muramatsu KM, Ramirez M. Coordination in humanitarian relief chains: Practices, challenges and opportunities. International Journal of Production Economics. 2010;**126**(1):22-34

[4] Van den Berg JP. Highly Competitive Warehouse Management. Buren: Management Outlook Publising; 2012

[5] Bartholdi J, Hackman S. Warehouse and Distribution Science, the Supply Chain and Logistics Institute. Atlanta: School of Industrial and Systems Engineering; 2011

[6] Kembro JH, Norrman A, Eriksson E. Adapting warehouse operations and design to omni-channel logistics: A literature review and research agenda. International Journal of Physical Distribution & Logistics Management. 2018;**48**(9):890-912

[7] He M, Shen J, Wu X, Luo J. Logistics space: A literature review from the sustainability perspective. Sustainability. 2018;**10**(8):2815

[8] Mohamud IH, Kafi MA, Shahron SA, Zainuddin N, Musa S. The role of warehouse layout and operations in warehouse efficiency: A literature review. Journal Européen des Systèmes Automatisés. 2023;**56**(1): 61-68

[9] Greeff G, Ghoshal R. Practical E-Manufacturing and Supply Chain Management. Elsevier; 2004. Available from: https://shop.elsevier.com/books/practical-e-manufacturing-and-supply-chain-management/greeff/978-0-7506-6272-7

[10] Richards G. Warehouse Management: A Complete Guide to Improving Efficiency and Minimizing Costs in the Modern Warehouse. Kogan Page Publishers; 2017. Available from: http://www.koganpage.com/

[11] Goodhue DL, Wixom BH, Watson HJ. Realizing business benefits through CRM: Hitting the right target in the right way. MIS Quarterly Executive. 2002;**1**(2):79-94

[12] Škerlič S, Muha R, Sokolovskij E. Application of modern warehouse technology in the Slovenian automotive industry. Transport. 2017;**32**(4):415-425

[13] Zhen L, Li H. A literature review of smart warehouse operations management. Frontiers of Engineering Management. 2022;**9**(1):31-55. DOI: 10.1007/s42524-021-0178-9

[14] Kembro J, Norrman A. The transformation from manual to smart warehousing: An exploratory study with Swedish retailers. The International Journal of Logistics Management. 2022;**33**(5):107-135. DOI: 10.1108/IJLM-11-2021-0525

[15] Valentini G, Zavanella L. The consignment stock of inventories: Industrial case and performance analysis. International Journal of Production Economics. 2003;**81**:215-224

[16] Cui W, Yang Y, Di L, Dababneh F. Additive manufacturing-enabled supply chain: Modeling and case studies on local, integrated production-inventory-transportation structure. Additive Manufacturing. 2021;**48**:102471

[17] Stošić-Mihajlović L, Marković J. Organization of logistics and procurement chain and entrepreneurship in accordance with halal standard. Ekonomski izazovi. 2022;**11**(22):18-30

[18] Hamel D, Rozman V, Liška A. Storage of cereals in warehouses with or without pesticides. Insects. 2020;**11**(12):846

[19] Saderova J, Rosova A, Behunova A, Behun M, Sofranko M, Khouri S. Case study: The simulation modelling of selected activity in a warehouse operation. Wireless Networks. 2022;**28**(1):431-440

[20] Gu J, Goetschalckx M, McGinnis LF. Research on warehouse operation: A comprehensive review. European Journal of Operational Research. 2007;**177**(1):1-21

[21] Saleheen F, Miraz MH, Habib MM, Hanafi Z. Challenges of warehouse operations: A case study in retail supermarket. International Journal of Supply Chain Management (IJSCM). 2014;**3**(4):63-67

[22] Gunasekaran A, Marri H, Menci F. Improving the effectiveness of warehousing operations: A case study. Industrial Management & Data Systems. 1999;**99**(8):328-339

[23] Autry CW, Daugherty PJ. Warehouse operations employees: Linking person-organization fit, job satisfaction, and coping responses. Journal of Business Logistics. 2003;**24**(1):171-197

[24] Lim MK, Bahr W, Leung SC. RFID in the warehouse: A literature analysis (1995-2010) of its applications, benefits, challenges and future trends. International Journal of Production Economics. 2013;**145**(1):409-430

[25] Min H. The applications of warehouse management systems: An exploratory study. International Journal of Logistics: Research and Applications. 2006;**9**(2):111-126

[26] Khan MG, Huda NU, Zaman UKU. Smart warehouse management system: Architecture, real-time implementation and prototype design. Machines. 2022;**10**(2):150

[27] Ramaa A, Subramanya K, Rangaswamy T. Impact of warehouse management system in a supply chain. International Journal of Computer Applications. 2012;**54**(1):0975-8887

[28] Perkumienė D, Ratautaitė K, Pranskūnienė R. Innovative solutions and challenges for the improvement of storage processes. Sustainability. 2022;**14**(17):10616

[29] Andiyappillai N. Data analytics in warehouse management systems (WMS) implementations–a case study. International Journal of Computer Applications. 2019;**181**(47):14-17

[30] Torabizadeh M, Yusof NM, Ma'aram A, Shaharoun AM. Identifying sustainable warehouse management system indicators and proposing new weighting method. Journal of Cleaner Production. 2020;**248**:119190

[31] Caridade R, Pereira T, Ferreira, LP, Silva F, Analysis and optimisation of a

logistic warehouse in the automotive industry. Procedia Manufacturing, 2017;**13**:1096-1103

[32] Mostafa N, Hamdy W, Alawady H. Impacts of internet of things on supply chains: A framework for warehousing. Social Sciences. 2019;**8**(3):84

[33] Wang H, Chen S, Xie Y. An RFID-based digital warehouse management system in the tobacco industry: A case study. International Journal of Production Research. 2010;**48**(9):2513-2548

[34] Aulkemeier F, Paramartha MA, Iacob M-E, van Hillegersberg J. A pluggable service platform architecture for e-commerce. Information Systems and e-Business Management. 2016;**14**: 469-489

[35] Raj GA, Sampath MK, Venkatesh V. Opportunities for cloud based software as a service (SaaS) warehouse management system an Indian industry insight. SAMVAD. 2013;**6**(2):43-60

[36] Fosso Wamba S. Achieving supply chain integration using RFID technology: The case of emerging intelligent B-to-B e-commerce processes in a living laboratory. Business Process Management Journal. 2012;**18**(1):58-81

[37] Singh A, Narain R, Yadav RC. An exploratory study of the SCM practices and IT usage: An emerging market context. International Journal of Information Technology and Management. 2010;**9**(4):446-467

[38] Wang X, Kumar V, Kumari A, Kuzmin E. Impact of digital technology on supply chain efficiency in manufacturing industry. In: Digital Transformation in Industry: Digital Twins and New Business Models. Cham: Springer; 2022. pp. 347-371. DOI: 10.1007/978-3-030-94617-3_25

[39] Kmiecik M. Logistics coordination based on inventory management and transportation planning by third-party logistics (3PL). Sustainability. 2022;**14**(13):8134

[40] Lieb R, Kendrick S. The use of third-party logistics services by large American manufacturers, the 2002 survey. In Supply Chain Forum: An International Journal. Taylor & Francis. 2002;**3**(2):2-10. DOI: 10.1080/16258312.200211517100

[41] Lieb RC, Bentz BA. The use of third-party logistics services by large American manufacturers: The 2003 survey. Transportation Journal. 2004;**43**(3):24-33

[42] Kamali A. Smart warehouse vs. traditional warehouse. CiiT International Journal of Automation and Autonomous System. 2019;**11**(1):9-16

[43] Abdalla S, Apramian SS, Cantley LF, Cullen MR, Mock C, Nugent R, et al. Occupation and risk for injuries. In: Injury Prevention and Environmental Health. 3rd ed. The International Bank for Reconstruction and Development/ The World Bank; 2017. PMID: 30212110

[44] Weidinger F, Boysen N, Schneider M. Picker routing in the mixed-shelves warehouses of e-commerce retailers. European Journal of Operational Research. 2019;**274**(2):501-515

[45] Gutelius B, Theodore N. The Future of Warehouse Work: Technological Change in the US Logistics Industry. UC Berkeley Labor Center. Working Partnerships USA; October 2019. Available from: https://laborcenter.berkeley.edu/future-of-warehouse-work/

[46] Van Geest M, Tekinerdogan B, Catal C. Design of a reference architecture for developing smart warehouses in industry 4.0. Computers in Industry. 2021a;**124**:103343

[47] Van Geest M, Tekinerdogan B, Catal C. Smart warehouses: Rationale, challenges and solution directions. Applied Sciences. 2021b;**12**(1):219

[48] Chow HK, Choy KL, Lee WB, Lau K. Design of a RFID case-based resource management system for warehouse operations. Expert Systems with Applications. 2006;**30**(4):561-576

[49] Dukic G, Opetuk T. Warehouse layouts. In: Manzini Warehousing Warehousing in the global supply chain: Advanced models. Tools and Applications for Storage Systems. 2012:55-69. DOI: 10.1007/978-1-4471-2274-6

[50] De Koster RB, Johnson AL, Roy D. Warehouse Design and Management. Taylor & Francis. 2017;**55**(21):6327-6330. DOI: 10.1080/00207543.2017.1371856

[51] Pane SF, Awangga RM, Azhari BR, Tartila GR. RFID-based conveyor belt for improve warehouse operations. TELKOMNIKA (Telecommunication Computing Electronics and Control). 2019;**17**(2):794-800

[52] Luo L, Zhao N, Zhu Y, Sun Y. A* guiding DQN algorithm for automated guided vehicle pathfinding problem of robotic mobile fulfillment systems. Computers & Industrial Engineering. 2023;**178**:109112

[53] Yoshitake H, Kamoshida R, Nagashima Y. New automated guided vehicle system using real-time holonic scheduling for warehouse picking. IEEE Robotics and Automation Letters. 2019;**4**(2):1045-1052

[54] Tong Q, Ming X, Zhang X. The realization for automated warehouse based on the integration of ERP and WMS. In: Paper presented at the Proceedings of the 2020 the 7th International Conference on Automation and Logistics (ICAL). New York, USA; 2020:76-80. DOI: 10.1145/3412953.3412954

[55] Lee CK, Lv Y, Ng K, Ho W, Choy KL. Design and application of internet of things-based warehouse management system for smart logistics. International Journal of Production Research. 2018;**56**(8):2753-2768

[56] Su X, Chu C-C, Prabhu B, Gadh R. On the creation of automatic identification and data capture infrastructure via RFID. In: The Internet of Things: From RFID to the Next-Generation Pervasive Networked Systems. Auerbach; 2007. Available from: https://www.routledge.com/The-Internet-of-Things-From-RFID-to-the-Next-Generation-Pervasive-Networked/Yan-Zhang-Yang-Ning/p/book/9780429133152

[57] Goli S, McKane A, Olsen D. Demand response opportunities in industrial refrigerated warehouses in california. 2011. Retrieved from

[58] Roodbergen KJ, Vis IF. A survey of literature on automated storage and retrieval systems. European Journal of Operational Research. 2009;**194**(2):343-362

[59] Azadeh K, De Koster R, Roy D. Robotized and automated warehouse systems: Review and recent developments. Transportation Science. 2019;**53**(4):917-945

[60] Custodio L, Machado R. Flexible automated warehouse: A literature review and an innovative framework. The International Journal of Advanced Manufacturing Technology. 2020;**106**:533-558